Mother Nature's Two Laws:

Ringmasters for Circus Earth

Mother Nature's Two Laws:

Ringmasters for Circus Earth

Lessons on Entropy, Energy, Critical Thinking, and the Practice of Science

A.D. KIRWAN, JR.

Graduate College of Marine Studies
Universty of Delaware

Published by

World Scientific Publishing Co. Pte. Ltd.

P O Box 128, Farrer Road, Singapore 912805

USA office: Suite 1B, 1060 Main Street, River Edge, NJ 07661

UK office: 57 Shelton Street, Covent Garden, London WC2H 9HE

British Library Cataloguing-in-Publication Data
A catalogue record for this book is available from the British Library.

MOTHER NATURE'S TWO LAWS: RINGMASTERS FOR CIRCUS EARTH

ISBN 981-02-4314-6

Printed in Singapore.

This book is dedicated to Dede Kirwan,
who cheerfully slogged through
many versions of the manuscript.
Her insistence on clarity, conciseness,
and critical objectivity provided invaluable guidance,
which, unfortunately, I was not always able to follow.

I also dedicate this book to our sons,
Michael, Patrick, and Ab.
You four have taught me the important lessons of life.

Foreword

You, the public, are increasingly aware of the importance of science and technology to decisions in modern life as well as to world survival. Yet, most nonscientists have had a limited exposure to the underlying principles that govern the everyday operations of nature. Faced with the deluge of information in today's media proposing to solve complex problems, many of us are beginning to feel like passengers on an airplane voting on landing instructions for the pilot.

Technology relies on science and science relies on observation, theory and experiment. For the last two centuries, humans have benefited from and usefully exploited two fundamental laws of nature, the first and second laws of thermodynamics. These laws, or fundamental theories, are in accord with scientific experiment and observation. To this day, despite many attempts, no one has found a single violation. Professor Kirwan calls these laws Mother Nature's Two Laws.

The book is divided into two sections. The first teaches the basics of the two fundamental laws in terms accessible to the reader. This discourse is designed to develop in the reader both an understanding of these laws and a sense of their application in practical terms.

The first law, conservation of energy, informs us that, while we can convert energy between its different forms, the total amount of energy remains constant. The second law shows how heat energy may be converted to useful work energy. Empowered with the second law, humans have created machines from steam engines to space stations to computer networks; and are continuing to do so at a dizzying rate. But the second law is not only about how to convert energy to useful work, it also quantifies exactly the cost of this conversion. That cost is an ever-decreasing amount of "useful" energy and is measured by the increase in a quantity called entropy. Entropy and its necessary increase are not readily intuitive, but the concepts are well established and have been confirmed repeatedly by scientific observation.

Today with the enormous growth in population and the attendant expansion in the exploitation of energy and material resources, we are faced with concerns regarding the sustainability of humankind's continued development on Earth. These concerns, when translated into political debate, place the citizen, whether or not a scientist, in the position of evaluating the arguments made. This book conveys two important messages regarding arguments purportedly based on science or technology. First, leaders, whether political, commercial or journalistic, should rigorously review their comments to ensure that they are not in violation of basic natural law. Second, the individuals receiving such comments should be very skeptical of sources that lack appropriate review.

But how do you review arguments regarding complex issues? In the second half of the book, Kirwan provides some useful tools and applies them to a variety of problems, both large and small.

The first tool Kirwan introduces is the scientific protocol, which has had continuous success in addressing technical issues. This tool emphasizes the taking of careful observations followed by peer review of the observations themselves and of the conclusions drawn from them. The second tool is a set of "ingredients" or guidelines for critical thinking. Both tools, when applied to complex issues, provide the basis for rational discussion, debate, and decision.

To give insight into how to apply scientific protocol, critical thinking and Mother Nature's Two Laws, Kirwan takes the reader through several detailed examples. Some, such as the attendance of over 100,000 people at the 50th anniversary of the purported alien landing in New Mexico and the panic over the New Madrid earthquake predictions, describe relatively local situations where media and enthusiasts exploited excitement and fear. Other problems, such as global warming, industrial waste, and energy supply have strong national and international implications. Such issues are complex, daunting, and the subject of continued debate. For these issues, it is especially important that incorrect or intentionally misleading arguments not be allowed to triumph over critical review and scientific debate.

By following Kirwan's approach, readers come away with a perspective so often lacking in less organized and more verbose discussions. The author demonstrates how a systematic approach and critical thinking, combined with a basic knowledge of the laws of science, can be a most useful guide for individuals interested in a wide range of scientific problems. While successfully accomplishing this, Mother Nature's Two Laws: Ringmasters for Circus Earth, does more. In addition to informing the interested layman, the author takes careful aim at the responsibilities of those "special" individuals who shape public opinion, whether as journalists, politicians, academic experts, religious leaders or teachers. He exhorts these individuals to endeavor to the extent possible within the ethic of their profession to ensure that their activities recognize the importance of careful review and critical thinking. It is harmful and sometimes dangerous for individuals, respected by the public because of their positions, to report or publicize scientific claims without a careful investigation of their validity. This statement applies however interesting or newsworthy these claims may be or however well they might support a position or belief.

Critical thinking weaves between science and politics and is not captured well in a five-second sound bite. Nevertheless, there is nothing to prevent its application prior to widespread dissemination of science-based opinion. Our leaders would do well if they applied Professor Kirwan's ideas to both their thinking and public statements.

W. J. Merrell
Senior Fellow and President
The H. John Heinz III Center for Science, Economics and the Environment

Preface

"A good many times I have been present at gatherings of people who by standards of the traditional culture are thought highly educated and who have with considerable gusto been expressing their incredulity at the illiteracy of scientists. Once or twice I have been provoked and have asked the company how many of them could describe the Second Law of Thermodynamics. The response was cold: it was also negative. Yet I was asking something which is about the scientific equivalent of: have you read a work of Shakespeare's?"

C. P. Snow

The law Professor Snow referred to, along with the First Law of Thermodynamics, comprise the two most fundamental principles in all of science. Scientists universally refer to these principles as the First and Second Laws. They are so fundamental to everyday experience that I call them Mother Nature's Two Laws or MN2L. The First Law presumes the existence of an ethereal quantity called *energy*, which has many forms or faces. This law states that the amount of energy doesn't change when it is transformed from one face to another. The Second Law presumes the existence of another ethereal, multi-faced quantity—*entropy*. This law states that the transformation of energy to work, that is, something useful for humankind, is never 100% efficient. The measure of this inefficiency is entropy. These two laws are one of the focuses of this book.

Professor Snow made his observation in 1959 in *The Two Cultures and the Scientific Revolution*. The intelligentsia he was referring to were essentially other academics and their disciples, many of whom were educated in the liberal arts and humanities. When I read Professor Snow's book shortly after taking my first academic position at New York University, I dismissed this "cultural gap" as just typical academic one-upmanship. I could not have been more wrong. The cultural bifurcation he referred to has now spread beyond academia into the mainstream media and politics.

I came to this realization in the early 1980's when a network news anchor devoted a segment of the nightly news to a report of a strange device. According to the news report, the device, when plugged into a regular wall outlet, delivered more power than was supplied through the electrical circuit. As an incorrigible "news junkie," I found this report especially distressing. Was the national news media so uninformed as to consider an obvious violation of the First Law worthy of coverage? Could it be that news editors, presumably college graduates with considerable training and experience in writing, were actually the scientific equivalents of illiterates?

Unfortunately, this was not an isolated incident. Subsequently, I have amassed a substantial file of misstatements, curious legislation, and other assorted actions culled from the media and political arenas regarding science and technology. Many of the examples are so silly or inane that our Neolithic ancestors would have been amazed and amused at our ignorance.

This file is the basis for two premises. First, virtually all information the public gets on developments in science and technology comes from the news media and the legislative and executive branches of the federal government. Second, these sources are not reliable.

Perhaps not coincidentally, there is a rising chorus of antipathy toward science and technology. This attitude has always been popular with fundamentalist and disenfranchised groups, but in recent years, many of the "establishment" have joined in the chorus. In some sophisticated circles, it is now fashionable to flaunt ignorance of science and technology. Ironically, many of these groups have acquired considerable technical expertise to push their agendas.

As noted earlier, the news media often treat topical issues in science and technology with cavalier disdain. A common rationale is that "science doesn't sell news." This is surprising since most scientists I know are impressed by the public's appetite for science. Stephen Hawking's books have appeared on *The New York Times* best sellers list; there seems to be an unlimited demand for popular books on relativity; and the creator of this theory, Albert Einstein, was one of the most widely recognized and respected figures of the 20th century. More recently *chaos* has caught the public's fancy, as evidenced by James Gleick's popular book by this name.

There is also a disturbing counter-theme to the present disdain for science and technology. Beginning with the Industrial Revolution, the myth has evolved that, through science and technology, humankind is "conquering" or "taming" nature. Policy based on this myth not only runs the risk of being wasteful and ineffective, it can be dangerous. Economic globalization is a consequence of the marvelous scientific and technical advances made during the past 200 years. However, these advances have not freed us from the whims of Mother Nature; they have made us more dependent than ever. Natural events such as hurricanes, droughts, floods, earthquakes, volcanic eruptions, and solar flares now have worldwide economic consequences.

In Shakespeare's time, all the world was a stage. Today, The Bard might say that it resembles a farcical opera. On one side of the stage, the anti-science/technology chorus is blaring forth discordant crescendos, while on the other side, an equally clamorous chorus proclaims magical "scientific" elixirs for all of life's problems. Scurrying between the two are pompous couriers. Periodically they stop to make comically absurd pronouncements about what the two choruses are singing.

For most of us life is not a comic opera. The impact of science and technology is real and rapidly accelerating. So how can the typical person make any sense out of the cacophony? Much of the hyperbole can be unmasked by a modest amount of critical thinking, an appreciation of the way science is practiced, and a qualitative understanding of the two most sacred principles in science, the First and Second Laws of Thermodynamics.

Critical thinking and the practice of science seem to have broad appeal. We are beginning to be submerged by a tidal wave of books with titles akin to "Critical Thinking for the Soul." The practice of science goes far beyond the scientific method but does not require technical training to appreciate. Moreover, the public is interested in learning about the hoops scientists must jump through before their research is acknowledged by their peers.

But why thermodynamics? At a basic level, everything behaves as a thermodynamic system. Energy flows in, this is transformed to something useful, and there is the inevitable generation of entropy that is expelled in waste products. Hence, everyone needs to have at least some qualitative understanding of the First and Second Laws of Thermodynamics.

Of course, there is much more to science and technology than these laws. Principles such as Newtonian dynamics, electromagnetic theory, and quantum mechanics have made well-documented contributions to the modern lifestyle. Substantial scientific training is required to appreciate the significance these principles have had on society. For many not trained in science and technology, this knowledge gap is overwhelming and the general tendency is to simply give up trying to learn anything scientific or technical.

But MN2L have the widest application of any scientific concept. To understand how a radio works requires a background in electromagnetic theory and solid state physics. Similarly, understanding how the human body works requires considerable background in biology. But radios and people share one thing in common: Energy flows into both, each performs a function or group of functions that has less energy value than the input energy, and each generates waste products. Regardless of their functions, at the most elemental level both are thermodynamic systems.

The same is true of a petunia and a power plant or a snail and a 747. Regardless of their relative sizes or disparate functions; each takes in energy in the form of fuel, food, or photons; performs mandated functions; and expels waste material. As elemental thermodynamic systems all conform to MN2L. It follows that knowledge of the basics of thermodynamics is the first step in understanding our world.

As noted so eloquently by Professor Snow in his book, qualitative understanding of the First and Second Laws is part of the human cultural heritage. Since they are accessible at a qualitative level and have universal application, these two laws represent the minimum amount of scientific knowledge nonscientists need today. Perhaps exposure to the First and Second Laws will stimulate readers to learn about other principles of science.

MN2L differ from all other scientific laws in that they are scale and discipline independent. That is, they apply to phenomena ranging from cosmic to quantum scales. Only a few scientific and technical disciplines use quantum mechanics or the theory of relativity routinely, but all disciplines insist on conformity to the First and Second Laws. Moreover, MN2L are interdependent and generally must be used together. Finally, it is difficult to imagine quantum theory being applicable to economics or art, yet these are just two of several nonscientific fields in which concepts arising from the First and Second Laws have had some useful applications.

This book is intended for those who are curious about nature, but have little scientific and technical training. My approach is different in at least three ways from other books of this genre. First, I make no attempt to provide a broad survey. Instead, my goal is to provide nonscientists with a qualitative understanding of the two most fundamental principles of nature and to provide some examples of how they affect our lives. To do this, I have also taken a different approach than what might be expected of a book on science and thermodynamics. I have tried to implement Professor Snow's thesis that MN2L are really part of our cultural heritage. Hence, the emphasis is on generic systems rather than traditional thermodynamic systems such as engines and refrigerators. This allows readers to look at everyday objects or even economics and the environment from a thermodynamic perspective. The common thread linking all systems is that they require some form of energy from their environments to perform assigned tasks. In doing so, they generate wastes that must be transferred back to their environments.

Another difference is that I emphasize the practice of science, or what I call the scientific protocol. The three pillars of the scientific protocol are: (1) the use of the scientific method in research; (2) the free exchange of data; and (3) the publication of research results in peer-reviewed journals. Most readers have had a cursory exposure to the scientific method, but the importance of peer-reviewed publications and free exchange of data generally are hardly mentioned in obligatory science for nonscientists courses, or in popular science books by prominent scientists. Therefore, many readers have only a miniscule understanding of the protocol, its importance in science, and hence indirectly, its effect on society.

The final difference is that I spend some time on critical thinking. Here, I have been influenced by Professor A. B. Arons who, in my view, best explains this topic in his book, *Teaching Introductory Physics*. As Arons states, this is a skill that requires considerable practice in more than one field. Armed with a qualitative understanding of MN2L, the scientific protocol, and exposure to critical thinking, readers will be less vulnerable to bogus scientific claims and not oblivious to obvious dangers from ill-conceived applications of science and technology.

The book is organized into two parts. The goal of Part 1, Mother Nature's Two Laws, is to explain MN2L in simple, qualitative terms. Chapter 1 gives an elementary explanation of these laws along with a bit of historical perspective. Chapter 2 describes the First Law, the conservation of energy, in detail. I offer an extensive catalog of different types of energy along with an energy gauge for the work per unit mass that can be obtained from each type. This chapter closes with a historical perspective of the societal impact of some of the types of energy. Chapter 3 focuses on the Second Law, the inevitable generation of entropy whenever energy is transformed to work. Here, I discuss the statistical and classical faces of entropy. An important theorem on minimum entropy production, proposed by Ilya Prigogine, is also introduced. The chapter concludes with a discussion of the connections between entropy and change.

The goal of the second part of the book, Ringmasters for Circus Earth, is to apply qualitatively MN2L, critical thinking, and knowledge of the scientific protocol to some topical issues. Chapter 1 introduces the scientific protocol and critical thinking. I advance the thesis that the evolution of the protocol, the recognition of the universality of MN2L, and the Industrial Revolution were interconnected and interdependent. One aspect of the scientific method is hypothesis testing, so a section of this chapter is devoted to this topic. The rest of the chapter presents Professor A. B. Arons's 10 tenets of critical thinking plus a couple of my own. Chapter 2 deals primarily with environmental and economic issues. The first part of this chapter addresses some global aspects of the conversion of energy to work and the consequent generation of entropy. The topical issues, the finite reservoir of hydrocarbon-based fuels and the never-ceasing search for clean energy, are discussed from the standpoint of the First and Second Laws. The balance of this chapter addresses the roles of MN2L in the environment and economics. The view adopted here is that industrial wastes and pollution are the agents that transport entropy away from industrial systems. I argue that the concept of "industrial ecology," as put forth by Robert Frosch of Harvard University, is one important way of establishing minimum entropy production for a collection of different industries. This has several potentially important implications for minimizing waste and pollution while simultaneously increasing profits.

Chapter 3 addresses global warming. First, I review some relevant scientific and technical issues. Next, I appeal to the principles of critical thinking to assess current political and media rhetoric about global warming. After this, I evoke a statement on global change by the American Geophysical Union; the world's largest scientific society of researchers studying basic geophysical processes related to the planet and its environment in space. Readers will note the position taken here is in sharp contrast to the claims by both global warming advocates and opponents that there is some scientific consensus on this issue. The last chapter, Chapter 4, addresses the roles of science in law, politics, journalism, and society.

As much of the material in Part 2 is concerned with political hot button topics, a caveat is in order. Although scientists may have political views, science and technology are apolitical. Most scientists are able to separate their politics from their science. In fact, one characteristic of successful scientists is their ability to suppress personal biases in arriving at interpretations of observations. This ability is also an important characteristic of critical thinking.

The book is written for nonscientists. The discussions and examples are greatly simplified; however, I make an effort to expose readers to some cutting-edge scientific concepts and issues. Although the focus is on the First and Second Laws, a number of asides touching on related scientific issues are included as well. I attempt to alert the reader to those topics that are subjects of current research and, thus, controversial. The use of mathematics, the natural language of science, is kept to the level of high school algebra. While there are very few citations in the text, each chapter has annotated suggestions for additional reading for those interested in more in-depth discussion of points covered in the text. Most of these are tailored for nonscientists; however, citations to a number of articles from *Science* are included. Nonscientists should find these articles informative. There are also recommendations for a few technical articles by some of the best scientists. First-rate scientists also happen to be extraordinarily gifted writers. I believe the introductions and conclusions to some of their technical articles are the best sources of overviews. Nonscientists should benefit greatly from even a modest exposure to cutting-edge science as it appears in the scientific literature.

Many readers will not want to read this book straight through. Rather, they might benefit by first reading those sections of most interest to them. For example, some readers might go directly to material in Part 2, others might prefer to read Part 1 and omit Part 2, and still others might wish to read Chapter 1 of Part 1 and then skip directly to Part 2. I have tried to put in enough road signs so that those readers who are skimming through the book will be able to detour to sections for appropriate technical background or related material.

A comment on the title of the book is appropriate. This was inspired by a colleague, Bill Seitz of Texas A&M University, who once wrote:

> "Science, art, literature, commerce, and engineering seek to understand, describe and/or use people, plants, animals, energy, materials, oceans, the atmosphere, and space. As individuals, we are both participants and spectators in the resulting circus. This analogy has been made many times before, but only recently have we come to realize the circus tent is sealed.

We play dual roles in the circus. In our role as spectators we have come to expect ever-grander scientific and technological feats to provide us with continuous amusement, excitement, instant gratification, and eternal life. This expectation is attributed to the scientific and technological advances of the past 200 years. However, during this time humanists, philosophers, and artists have come to attack the increasingly exotic spectacle as we race toward a seemingly ever-expanding future inside the tent. In our role as circus performers we realize there are limits to such behavior."

Where does this sense of limitation come from? That is the subject of this book.

Acknowledgements

Some of the ideas expressed here about the application of general thermodynamic concepts to environmental issues were first presented in my 1996 talk at the retirement ceremony for Professor William Sackett at the University of South Florida. The extension of thermodynamic reasoning to ecological systems was made in a lecture Fred Dobbs and I jointly gave in 1997 at Old Dominion University. The bulk of the work on this manuscript was completed while I was on sabbatical at the University of Wyoming's Department of Mechanical Engineering during the spring term of 1998. I am indebted to Dean Kynric Pell of the College of Engineering, Department Chairman William Lindberg, and Professor Carl Reid for their gracious hospitality. The congeniality and intellectual atmosphere in the department and college were most stimulating. I am especially indebted to my department host, Professor Andy Hanson. His assistance and support made the sabbatical both productive and enjoyable. I also am pleased to acknowledge the support provided by Old Dominion University during the sabbatical. The book was completed after I joined the faculty of the University of Delaware. I thank Dean Carolyn Thoroughgood of the College of Marine Studies for providing the resources that allowed me to complete this effort.

Michael, Patrick, and Ab Kirwan gave valuable critiques. Pam Donnelly and Dan Lane provided very professional help in preparation of the figures and proofreading the manuscript. Lane McLaughlin was responsible for the cover design. Karal Gregory helped immeasurably in rewriting and editing. Bruce Lipphardt made a constructive review of Part 1. Bill Seitz provided an exceptionally thorough scientific review of an early draft and offered a number of very helpful suggestions, which I have attempted to implement. Finally, I thank Carole Blett for critical help in completing the manuscript. Both Carole and Karal refused to let the manuscript drop off my radar horizon.

I have tried to credit all who hold copyright or reproduction rights for quotations and figures reproduced here. These include:

- William Seitz for permission to use the unpublished quote that appeared in this preface.

- John Wiley & Sons for the tenets of critical thinking given in Chapter 1, Part 2. These tenets were taken from *Teaching Introductory Physics* by A. B. Arons.

- Princeton University Press for Figure 2.1, "Hubble pimple" and for the quote by H. Holland and U. Peterson at the start of Chapter 2, Part 2.

- *Science* for the quote taken from the article, "Economic Growth, Carrying Capacity, and the Environment," that appeared in Volume 268, 1995.

- The American Institute of Physics for three figures from "Industrial Ecology: Minimizing the Impact of Industrial Waste" that appeared in *Physics Today*, Volume 47, 1994.

- Robert A. Frosch, the author of "Industrial Ecology: Minimizing the Impact of Industrial Waste," for permission to use three figures from that article.

- American Geophysical Union for the statement in Chapter 3, Part 2 on Climate Change and Greenhouse Gases.

Contents

List of Tables and Figures

Part 1 Mother Nature's Two Laws

Part 2 Circus Earth

Part I List of Tables and Figures

Part 1 [illegible] Nature's Two Cases

Part 2 Circus Earth

Part 1 Mother Nature's Two Laws

1. Mother Nature's Two Laws: What Are They and Why Must We Heed Them?

"A theory is more impressive the greater the simplicity of its premise, the more different kinds of things it relates and the more extended is its area of applicability. Therefore the deep impression which classical thermodynamics made upon me. It is the only physical theory of universal content concerning which I am convinced that, within the framework of the applicability of its basic concepts, it will never be overthrown."

A. Einstein

Preview

We are constantly bombarded with a bewildering array of "laws." Some are heuristic, such as the law of supply and demand, and some, such as "The Peter Principle" or "Murphy's Law," are anecdotal. Scientific laws, often named for their discoverers, are precise and technical in nature. Usually these laws are merely useful approximations of more general principles. Even more confusing, "law" sometimes is used loosely to refer to a hypothesis still under evaluation before canonization as a basic principle of nature. Science is not a democracy; there is no duly constituted congress of scientists that enacts scientific laws. Science is not a religion; laws are not divinely prescribed, everlasting, and exempted from experimental testing. Scientific laws and theories have a finite shelf life. They are the products of an exacting and lengthy process known as the scientific protocol.

Well-publicized theories, such as evolution and relativity, were first proposed as common explanations of a diversity of observations in biology and physics. Both have been under scientific scrutiny ever since. As I understand it, the scientific debates about evolution are concerned with extending Charles Darwin's ideas to explain why there are physiological differences in individuals of a species and to reconcile natural selection with genetics. No doubt these debates will produce further modifications and extensions of Darwin's original ideas. On the other hand, Einstein's theory of relativity has met every scientific challenge for nearly a century.

In contrast, the concern here is with the two most basic principles of nature. The Preface referred to them as Mother Nature's Two Laws, or MN2L for short. These two principles are no longer subject to observational or experimental verification. Rather, they often are used as the first test of hypotheses. The wide application of these principles in all scientific disciplines sets them apart from all other scientific laws.

When it comes to precedence, scientists can be very aggressive. This is one reason why many scientific laws are named after their discoverers. In part, because they are so basic, this has not been the case with MN2L. Many eminent scientists made important contributions to the quantification of MN2L. Five, whose contributions are noted in this book, are Sardi Carnot, Rudolph Clausius, J. P. Joule, Lord Kelvin, and Ludwig Boltzmann. Carnot's work predates that of the other four; however, their contributions were necessary before the fundamental importance of Carnot's research was appreciated.

MN2L arose out of efforts to develop efficient engines nearly 200 years ago. We now realize these principles govern virtually all natural phenomena. They quantify the intuitive notions that energy and mass are neither created nor destroyed, and that the inevitable consequence of useful work is the generation of entropy.

A simple qualitative statement of the First Law is:

The total mass and energy of the universe doesn't change.

As I will show in the next chapter, energy exists in different forms or faces. All life and economics exist to convert energy from one face to another. Mother Nature has given us a wonderful gift, the First Law, which governs the transformation of energy to its various faces. This law states that merely changing the face of energy does not change the amount of energy. Just as a financial audit must show that assets equal liabilities, Mother Nature requires an exact balance of energy and mass for any system and its environment.

However, She imposes a tax on carefully contrived (as in engines) conversions to a peculiar form of energy, which I call work. A traditional example of work that readers may have run across in high school science is the energy output of engines. Here, work is used in a broader sense to denote energy used to accomplish specific purposes. This includes both traditional examples as well as esoteric examples such as growth of organisms and even the production and sale of goods and services.

The Second Law is the tax Mother Nature imposes on the conversion of energy to work. A simple statement of the Second Law is:

The energy output of work is always less than the energy transformed to accomplish it.

Application of the First Law to everyday life requires clear scientific definitions of energy and work. Similarly, the application of the Second Law introduces another word, entropy. Entropy represents the fundamental notion of the inevitable inefficiency that occurs when energy is transformed to work. As with energy, there are different forms or faces of entropy.

Less appreciated and understood is the fact that inefficiency implicit in the Second Law is a fundamental tenet of nature and not the result of poor engineering, inept management, or corrupt politics. Of course, when such human factors are present, they will conspire to exacerbate the inefficiency imposed by Mother Nature. Whenever the limitations implied by the Second Law are ignored, expect folly to follow. And because this law is so fundamental and ubiquitous, everyone should have at least an intuitive understanding of its consequences. Perhaps this is what led Professor Snow to observe that ignorance of the Second Law is equivalent to ignorance of Shakespeare.

The quantification of MN2L is less than 200 years old. Certainly this was a great intellectual achievement that has produced vast societal consequences. However, it should be kept in mind that we did not discover these laws, we merely quantified them. All life forms have some innate grasp of these principles. Plants evoke the First Law by using the radiant energy of the sun along with carbon dioxide from the atmosphere to grow and procreate. The Second Law insures that they are not perfectly efficient in the use of the sun's energy, so there are waste products, such as oxygen. Animals convert the chemical energy in plants (and to a lesser extent in other animals) along with oxygen from the atmosphere for the same purposes. One of the metabolic waste products of animals is carbon dioxide.

The lifestyles of early humans probably were no different from that of other animals. Their daily existence revolved around hunting and gathering food for the energy necessary for growth and procreation. The hunter-gatherer lifestyle changed dramatically in the Neolithic era when humans learned to control fire, domesticate animals, and farm. This provided access to levels of energy on demand far beyond that required for mere existence. In the Neolithic era, energy became a commodity that could be channeled to activities other than survival.

The modern, rather effete version of this fundamental concept is disposable income. I suggest that access to "disposable energy" is what first distinguished humans from other animals. Without this, philosophers, artists, poets, and musicians would not have evolved.

As attested by the vast amount of relics such as paintings, tools, and pottery discovered in recent years, we now know that Neolithic societies had extraordinarily complex social structures. It is also known that Neolithics farmed and used domesticated animals both for food and labor. Moreover, they left elaborate burial sites, and in many places in the world, they constructed elaborate monuments. Perhaps the most famous is Stonehenge, which even today is a popular tourist attraction. Construction of such artifacts required Neolithic peoples to solve formidable problems in mechanics. Moreover, there are indications that many of the monuments were designed (at least in part) to indicate the equinoxes and perhaps other astronomical phenomena. That they did this without a written statement of Newton's laws of motion or computers demonstrates a solid intuitive grasp of fundamental principles of mechanics and astronomy.

Neolithics also must have had a comparable understanding of MN2L. They established villages and towns that served as hubs for agriculture and trade. The engines for their economy were humans, beasts of burden, and ovens for pottery making. Those engines required fuel in the form of food, wood, peat, and perhaps animal fats and wastes, just as today's engines require oil, gas, or coal. The success of these early industrial enterprises suggests that these people realized the conversion of energy contained in fuels to work produced wastes that had to be dealt with. Of course, the amount of energy required for their engines was insignificant compared to the standard in the industrialized world. Consequently, the waste products of those days were small and locally confined. They relied on only Mother

Nature to recycle their wastes. One suspects, however, that even then pollution was a major societal problem.

The rest of the chapter provides some scientific background of the simple statements of MN2L given above. It concludes with an aside on two other laws of thermodynamics.

Vocabulary

In order for readers to understand MN2L, it is necessary to develop a feel for several basic concepts. As is often the case with fundamental concepts, it is difficult to describe any one without reference to the others. Consequently, we shall use some idealized examples for illustration. Consider a car engine. The purpose of the engine is to move the car so as to transport passengers. This requires energy, typically supplied by gasoline. A second example is a steel mill. The energy required to produce the steel comes from coal. Finally, consider an organism. Its purpose is to grow and procreate. The energy to do this comes either from the photons in solar radiation or from chemical energy obtained by eating other organisms. Of course other essentials are required if these systems are to function properly: nutrients for an organism, iron ore for the mill, and oxygen and motor oil for the car engine (to name but a few).

Readers can construct a simple input-output diagram to represent all three of these examples. Draw a box on a sheet of paper to represent any of these systems. The region outside the box represents the environment. Now draw two arrows, one going into and one coming out of the box. The input arrow represents energy or fuel flowing into the system from the environment, while the output arrow represents the system function or work performed by the system on the environment. Of course more realistic diagrams would have arrows to represent oxygen, nutrient, and ore inputs as well as waste products.

Two corporeal entities and one ethereal entity were identified in these examples. The corporeal entities are the system and its environment while the ethereal entity is energy, or work. Physicists, for example, think of energy in terms of basic processes such as might occur in collisions of elementary particles in accelerators. Engineers, on the other hand, think of energy as the output of an engine or machine. These views share a common theme: Energy is conserved in any of these processes. The energy before and after the collisions of elementary particles is the same, just as the energy in the fuel is the same as that in the work performed plus the amount lost to engine inefficiencies.

Here is a quirky view of energy. All economic activity requires energy, and, as everyone is reminded each month, it costs. So in a way, the money we spend on gasoline, heating, and air conditioning is not lost; it is converted to energy and to work. The equivalence of money and energy is a cornerstone of economics.

Work performed by systems typifies the energy transformed from fuel, which was obtained from the environment. Readers should think of work as the energy required to achieve some goal or purpose. Typical examples are mechanical processes such as moving objects or turning shafts. A more esoteric example would be the growth of organisms.

Being ethereal, energy must be conveyed by corporeal substances. Matter is one of two agents for transporting energy or work. Fuel and food are typical substances that contain energy. However, being ethereal, the energy in these substances cannot be isolated from the atoms and molecules that make up the substances. Engines or machines convert energy to work, but as with energy and fuel, it is not possible to isolate energy in the fuel or work in an engine.

The other mechanism for transporting energy is through discrete elementary packets of pure energy called photons. Photons are one of the true wonders of nature. They are formed whenever any substance radiates energy. Once emitted, they travel at the speed of light; when they strike any substance they transfer all of their energy to the substance and cease to exist. Some readers may have heard of the wave-particle duality of radiation. In that paradigm, electromagnetic signals, such as light, radio, or radar, propagate as waves but also behave as particles. Photons are the corporeal aspect of the duality.

Einstein found that the amount of energy packed into a photon is inversely proportional to the wavelength of the radiation; that is, the shorter the wavelength, the more energy a photon carries. The proportionality constant is the product of Planck's constant, named for famed scientist Max Planck, and the speed of light. Both Planck's constant and the speed of light are universal constants of nature. Intelligent beings in a distant galaxy would know of both. Their measurements of the numerical values of these two constants would agree with ours if both sets of measurements were converted to a common system of units.

As we will see in Chapter 2, there are various types of energy, and everyday matter contains all of them. Unfortunately, no engine or system can transform every type of energy to work. Instead, engines are designed to utilize one or two specific types of energies. A coal-fired furnace, for example, is unable to transform any nuclear energy contained in a lump of coal to work. Whatever the transformation mechanism, the First Law demands that the total energy be conserved. Except for nuclear energy, the amount of mass carrying the energy is also unchanged. With nuclear processes, Einstein's famous law, $E=MC^2$, relates the amount of mass lost to the amount of energy of the transformation. This was a stupendous intellectual achievement in that it established an equivalency between energy and mass at the most fundamental level. Because of this equivalency, the general statement of the First Law is that mass and energy are conserved in any transformation.

Mother Nature's Second Law states that not all the energy released in the transformation of energy is available to do work. The notion of unavailable energy is a subtle but essential cornerstone of an industrialized society. As noted earlier, this concept was first quantified almost 200 years ago; however, primitive versions were established long before. A measure of the unavailability of energy is entropy. Entropy is one of the most profound concepts ever devised. Since its first use as a measure of the efficiency of engines, this concept has been extended to cover a bewildering array of scientific and technological topics. Economists, philosophers, and artists have also adopted this concept. Like energy, it is useful to think of entropy as a primitive, ethereal quantity carried by ordinary matter.

A major goal of science is to explain how systems work. The more learned about a system, the greater the tendency to break it down into a collection of subsystems. Subsystems, in turn, may be broken down into yet smaller subsystems. A system composed of a collection of interacting subsystems may be called a compound system. An example of a compound system is a car. It is constructed of subsystems such as an engine, drivetrain, air conditioning, tires, steering, a radio and CD, seats, various gauges to monitor the performance of critical subsystems, etc. All of these subsystems can be broken down even further. A radio is another convenient, everyday example of a compound system. Some of its component subsystems are an antenna, attendant controls to receive precise radio frequencies, and speakers. Although all of these component subsystems were designed to utilize other principles of science to accomplish specific functions, they all behave as thermodynamic systems and thus conform to MN2L.

Continued partitioning of the subsystems reveals that all of the materials making up the many subsystems of a car are composed of a large variety of molecules. The molecules are composed of atoms, which are composed of electrons and nuclei. The latter are composed of smaller bits of matter called nucleons. These, in turn, are composed of the fundamental building blocks of matter, which are generically labeled *quarks*. The breakdown of everyday objects or "things" to quarks is described simply and elegantly by M. Y. Han in *Quarks and Gluons* (see Suggested Reading at the end of this chapter).

It is comforting to know that there is ultimate structure in nature and that there are basic building blocks—quarks—common to any system. However, it is not yet possible to explain the properties of larger systems, like cars and refrigerators, by quark processes. A testament to the universality of MN2L is that they apply equally to systems ranging from atoms to galaxies. Despite all our high technology tools and toys, we cannot escape the jurisdiction of these two laws.

In reality, nearly all systems are *open*. This means they can exchange both mass, which contains energy, and pure energy (in the form of heat, radiation, and fields such as gravity) with their environments. An idealized case is a *closed* system, which can exchange only energy contained in heat, radiation, or fields such as gravity, but no mass, with its environment.

The sealed circus tent analogy quoted in the Preface is a poetic way of saying that to a high degree of approximation, Earth is a closed system. It receives a miniscule amount of mass in the form of meteorites from the solar system, but it absorbs an enormous amount of radiant energy, in the form of photons, from the sun. Earth also radiates photons back to space but ejects only a tiny amount of mass. Finally, Earth is an excellent example of a compound system. Some of its important subsystems are the biosphere, lithosphere, atmosphere, hydrosphere, and cryosphere. Each of these subsystems can be partitioned into smaller and smaller subsystems.

An even more idealized case is an *isolated* system. Such systems have no communication with their environments. We can construct systems that are very good approximations to isolated systems for some period of time. But the Second Law still applies to these systems. They must age, and ultimately, no matter how well made, they will rupture and become open systems that will interact with their environments.

Often there is great flexibility in defining subsystems. Generally, the grouping of subsystems is dictated by the goals of the analysis. There is some advantage in limiting the number of subsystems in quantitative assessments of compound systems, so many units that have a common functionality may be grouped as one subsystem for some analyses. In other studies, it may be necessary to account for the subtle differences within elements of a functional grouping.

Figures 1.1 and 1.2 summarize the discussion in this subsection. The first figure, 1.1, shows a simple system. The system is *open* if the photon, conduction, and mass arrows are all open; that is, mass and energy are exchanged with the environment. If both input and output mass arrows are closed, then the system is *closed*. The *isolated* case arises when all inputs and outputs between the system and the environment are closed. The second figure, 1.2, depicts a closed compound system made up of a collection of four subsystems that are themselves simple open systems.

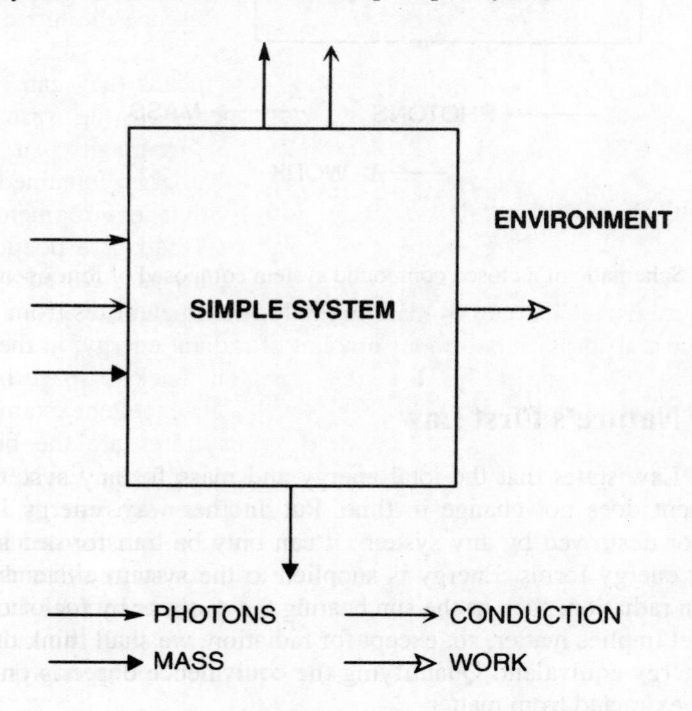

Fig. 1.1 Schematic of a simple system.

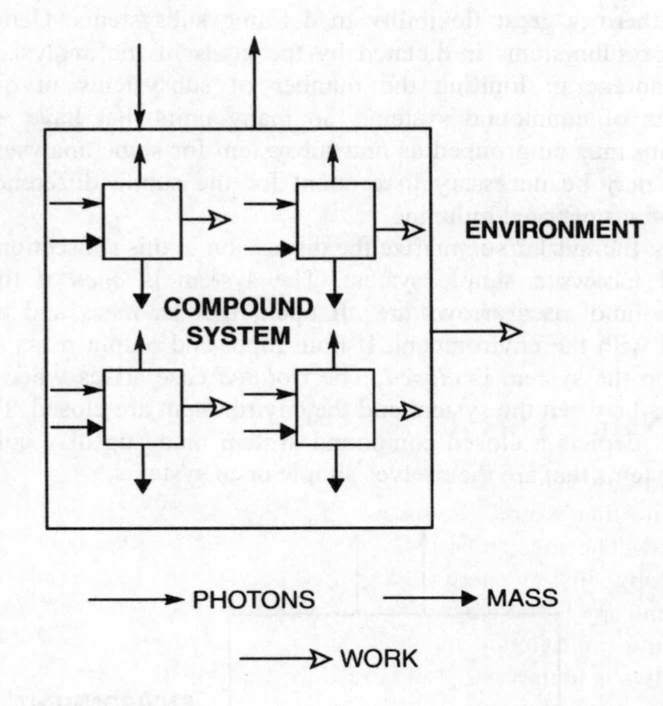

Fig. 1.2 Schematic of a closed compound system composed of four open simple systems.

Mother Nature's First Law

The First Law states that the total energy and mass for any system and its environment does not change in time. Put another way, energy is neither created nor destroyed by any system; it can only be transformed into work and other energy forms. Energy is supplied to the system either directly as photons in radiation, such as the sun heating the earth, or by fuel such as coal or oil. Fuel implies matter, so, except for radiation, we shall think of mass as just an energy equivalent. Quantifying the equivalence depends on the type of energy extracted from matter.

Systems usually are highly specialized and can only perform a restricted number of transformations on their fuel to produce work. For example, we acquire the energy to get up in the morning, to go to work, and to carry out typical human functions from the food we eat. Similarly, car engines convert the chemical energy in gasoline to the kinetic energy of the pistons that, in turn, propel the vehicle. Our digestive system performs the same function as a car engine guzzling gasoline. Obviously, we cannot assimilate the car engine's fuel, nor can a car run on breakfast cereal. However, a careful audit of all the atoms of the expulsions from either system, plus any accumulations of mass by the system, always equals the atoms contained in the fuel. Of course, the molecular structure of the fuel is altered in both cases by the transformation of energy to work.

Mother Nature's Second Law

The Second Law states that the conversion of energy to work is not perfectly efficient. In other words, the output of work is always less than the energy transformed. The measure of this inefficiency is another ethereal quantity called entropy. Just as energy is the ethereal quantity in fuel that is used to do work, entropy is the ethereal quantity in the residue of the spent fuel that measures the inefficiency in transforming energy to work. Unlike energy, which is always conserved, any system or process that transforms energy to work always generates entropy.

There is an important corollary to the Second Law that is generally recognized but rarely stated. The corporeal residue left from the transformation of the energy in fuel to work is poisonous to the system. Unless this residue is expelled, the system will die. If a car cannot vent the fumes resulting from combustion of fuel, the engine will shut down. If organisms cannot vent their metabolic waste products, they too will die, horribly. The expulsion of the by-products of energy conversion takes energy that further decreases the amount of work that can be obtained from a given increment of fuel.

Taking this further, expelled material containing entropy also carries with it forms of energy that can serve as fuel for other types of systems. Moreover, the entropy products of one system are not necessarily fatal to other types of systems. Ecosystems exist because there is a whole suite of organisms that thrive on processing the metabolic wastes of other organisms. The decomposers, in turn, serve as food for larger elements of the ecosystem. If this were not the case, life would not be possible. In our vocabulary, ecosystems are examples of compound thermodynamic systems.

It is important to keep in mind that both laws be used simultaneously in any thorough and complete analysis.

Let us see how MN2L really governs a workout at your health club. Your body is an open thermodynamic system, so anticipate that there will be mass flows in and out of your body during your workout. After stretching and warming up, you start a routine that consists of aerobic exercises, such as stair climbing or skiing, then move on to strength exercises with weights or resistance machines. In going through your workout routine, you are converting the chemical energy stored in your body to kinetic energy in your motions. This is converted into the kinetic energy of machines, and, if you are using free weights, the conversion of kinetic energy to potential energy by lifting the weights in the Earth's gravity field. Conversion of chemical energy to kinetic energy requires oxygen, more specifically O_2. This is extracted by your lungs from the air you breathe. In all of this activity, there is no creation or destruction of energy or mass; you are merely converting one form of energy to another, in conformance with the First Law. Note that this conversion does not involve a mass exchange. You only transfer an ethereal quantity—energy—to the machines or weights.

What role does the Second Law play in the workout? The conversion of chemical energy to kinetic energy is not 100% efficient. Another form of energy—heat—is also generated. This can have serious metabolic consequences, so your body gets rid of the heat by radiation and by sweat. In the conversion of chemical energy to kinetic energy, there are waste products. One obvious product is carbon dioxide, or the CO_2 that is exhaled. Your body expels solid waste combustion products later. The generation of these waste products is the inevitable consequence of the Second Law. They, along with sweat, carry away the second ethereal quantity: entropy. You might not want to ponder what would happen if these products were not expelled.

Finally, consider your body's mass balance during the workout. The only direct weight loss is through the expulsion of the waste products and sweat, which is a Second Law process. There is no mass involved in the transfer of chemical energy to kinetic energy or in the body's heat radiation. Of course, if you over drink or eat energy bars during the workout, you can actually gain weight!

Aside on the Zeroth and Third Laws

The original formulations of the First and Second Laws were for thermodynamics. They are half of the laws of thermodynamics that so many undergraduates in science and engineering love to hate. The other two laws, the Zeroth and Third Laws, have less consequences outside the bounds of thermodynamics than do the First and Second Laws. Other than the brief discussion given here, they are not considered further.

The Zeroth Law identified an absolute temperature as a fundamental property of nature, and provided a basis for its measurement. Temperature is a way of quantifying "hotness." Temperature is also a way of characterizing the kinetic energy of the molecules, atoms, or even more basic particles that make up all matter. The Zeroth Law was recognized around 1931, long after acceptance of the First and Second Laws. The absolute temperature scale is called the Kelvin scale. On this scale water freezes at 273.2 degrees and boils at 373.2 degrees under standard conditions. The range of temperatures that can be achieved is impressive. Temperatures within a small fraction of a degree of absolute zero are now routinely achieved. At the other extreme, temperatures may approach several thousand degrees Kelvin in a blast furnace, and several million degrees in bodies like the sun.

One popular form of the Third Law states that it is impossible to reduce the absolute temperature of any substance to zero. As temperature is a measure of the kinetic energy of the basic constituents of matter, this principle says that there must always be some motion associated with the internal structure of any substance. There is never rest at the subatomic level.

An alternate version of the Third Law may be more appropriate here. At extremely low temperatures the Third Law requires the entropy of a compound system composed of molecules, electrons, and other basic particles to tend to zero as the temperature approaches absolute zero. Based on this version, some scientists argue that the Third Law may be a limiting case of the Second Law.

Bill Seitz, of Texas A&M University, provided me with a concise summary of these laws. Seitz thinks of the First Law as Mother Nature's rule that we cannot get more work out of an engine than the energy in the fuel it consumes, that is, we can't win, but only break even, when we convert energy to work. The Second Law says we can't convert all the energy in fuel to work except at absolute zero temperature, that is, we can't break even unless we are at absolute zero. The Third Law says we can't get to absolute zero.

Table 1.1 summarizes the four laws of thermodynamics and also gives my cursory attempt to establish the approximate dates when they were formulated. I yield to historians of science to settle disputes on exact dates. The names and approximate dates of these laws are wonderfully paradoxical. The Second Law was formulated before the First and the Zeroth Law after both.

Law	Statement	Date
Zeroth	Establishes absolute temperature scale	1931
First	Conservation of energy and mass	1851
Second	Entropy increases with time	1824
Third	Absolute zero unattainable	1931?

Table 1.1 The Four Laws of Thermodynamics.

Suggested Reading

P. W. Atkins, *The Second Law*, W. H. Freeman (Scientific American Library), 1994. The historical notes on the laws of thermodynamics are taken from this marvelous book. I recommend it highly to both scientists and nonscientists.

R. Feynman, *The Character of Physical Law,* The M. I. T. Press, 1965. This little book is an excellent discussion of scientific laws. It comprises seven lectures given by Professor Feynman at the Messenger Lectures at Cornell in 1964. Anyone interested in science will enjoy reading the book and learn a good deal about scientific laws. The lectures were taped by BBC and aired by BBC-2. These tapes may be obtained from most large public and university libraries.

M. Y. Han, *Quarks and Gluons*, World Scientific, 1999. Anyone interested in the nature of matter and the scientific developments that led to the discovery of quarks should read this little book. It is written for nonscientists by one of the pioneers in particle physics. The book cover has a diagram that nicely depicts the decomposition of "things" to quarks.

2. The Faces of Energy

"It is important to realize that in physics today, we have no knowledge of what energy is. We do not have a picture that energy comes in little blobs of a definite amount. It is not that way. However, there are formulas for calculating some numerical quantity, and when we add it all together it gives . . . always the same number. It is an abstract thing in that it does not tell us the mechanism or reason for the various formula."

R. P. Feynman

Preview

"Energy" is one of the most popular and perhaps overused words in the English language. According to the *Oxford English Dictionary*, 2nd edition, the root comes from a Greek word meaning "work." Over time its use has broadened. According to the *OED* its usage is now broken down into seven definitions. Many of the definitions are synonymous with "force" and "power." A crosscheck of the latter two words will confirm this. On the other hand, in science, energy is a fundamental concept for which there is no ambiguity. Moreover, "force" and "power" are basic scientific concepts that are related to energy, but hardly synonyms of energy or each other.

Scientific distinctions between energy, force, and power are critical at this juncture. In order to understand them it is necessary to review briefly some high school science material. First on the agenda are dimensions and units. The only three dimensions of concern here are length, mass, and time. For convenience, I shall use L, M, and T to denote these dimensions respectively. Presumably readers have an intuitive sense of each of these quantities. Measurements of these quantities are based on an a priori selected system of units. The common system used by nonscientists in the USA is feet, pounds, and seconds. When appropriate, these units are replaced by equivalents in inches, yards, miles, ounces, tons, minutes, hours, days, etc.

First some bad news; we aren't going to use that system here. Instead, we will use a common scientific system, the meter, kilogram, and second, which is universally known as the MKS system. In the MKS system energy is measured in joules. Weight conscious readers might relate energy to calories, well, actually kilocalories. Diet recommendations give numbers that are called calories. However, diet calories are really kilocalories or thousands of calories as might be determined from the scientific definition of calorie. The usage is so common among dieters the kilo has been dropped. For comparison, one kilocalorie is about 4,200 joules, so a daily diet of 2,500 diet calories is about 10 million joules.

For completeness, force is expressed in newtons in the MKS system. This is just over 0.2 pounds of force. Finally, power in the MKS system is measured in watts, which is a joule per second. A kilowatt is about 1.3 horsepower. Presumably readers are comfortable with kilowatts since this is the basis for utility charges. So you may have been using the MKS system all along.

The above is the minimum readers need to be conversant with developments in science and technology. If you are still uncomfortable with the MKS system or would like to develop as intuitive a sense of MKS units as you have with feet, pounds, etc., the next section provides a more detailed introduction. After this, there is a discussion of different kinds of energy and energy transformations. Then there are asides on orders of magnitude and an energy gauge that compares the amount of energy associated with various processes in a kilogram of matter. The chapter concludes with a brief discussion of the history of the technology for transforming energy to work and its impact on civilization.

Aside on the MKS System

Now for some good news. It really isn't necessary to continually convert between the two systems if you follow a simple three-step prescription. First, determine your height in meters. The simplest thing to do is to use a yardstick that has inches and yards on one side and centimeters and meters on the other. If a meter stick is not handy, then you can calculate your height in meters by multiplying your height in inches by 0.0254. This will give your height in meters to two decimal places or to one centimeter if you round. Most adults are between 1.5 to 2 meters tall. Next, determine your weight in kilograms. If you don't have a scale that gives this then you can calculate your kilogram weight by dividing your pound weight by 2.2. Most adults weigh between 50 to 100 kilograms. The last step is to forget the conversion factors but remember your weight and height in the MKS system.

For those who worry about these sorts of things, you should know that both meters and kilograms are defined in terms of precise measurements or standards kept at the International Bureau of Weights and Measures at Sevres, France. An encyclopedia is a good place to look up further details. Also, strictly speaking, the weight measurement is the force from the Earth's gravity field that your body exerts on the scales. Since the gravitational acceleration is constant to a sufficient degree of accuracy for these purposes, the scales are calibrated so that the reading is numerically the same as your mass expressed in pounds.

Before tackling force, energy, and power, it may be appropriate to review some basic kinematics. We start with the concept of velocity. This is a subtle concept. Adults have developed a sense of speed from driving cars and flying. But the scientific notions of speed and an associated vector indicating the direction of movement is not intuitive. For purposes here, consider velocity as the rate at which a moving object's position changes with time. Velocity has the dimensions of length divided by time or L/T, and in the scientific units used here is reported as meters per second. A brisk walking or nominal jogging pace for most healthy adults is one body length every second. So someone 1.6 meters tall can walk with a velocity of 1.6 meters per second. Closely connected with velocity is momentum, another commonly used word. Scientifically, its usage is precise. Momentum is simply the mass of an object times its velocity or ML/T dimension-wise. So, if our hypothetical individual is walking with a velocity of 1.6 meters per second and this person weighs 65 kilograms, then his/her momentum will be 104 kilogram meters per second. Later we will use powers of 10, so to get you used to this, the momentum in this case would be reported as approximately 100 kilogram meters per second, rounding off four grams. In contrast, a pickup truck may weigh 1,000 kilograms. If it were moving at the same velocity, then its momentum would be about 1,600 kilogram meters per second or over 150 times more than that of the 65 kilogram adult moving at the same velocity. Obviously, in collisions, it is momentum and not just velocity that really counts. Now consider acceleration. This is the rate at which the velocity of the object is changing with time. Its dimensions are L/T^2. If the individual above is moving at 1.6 meters per second, and comes to a stop in one second, then she/he has decelerated 1.6 meters per second each second.

Force is a dynamical concept. It is quantified by the second best known formula in all science, F=MA. You can score trivia points if you remember this is actually the second of Newton's three laws. Here, F is the force, M is the mass of an object, and A is its acceleration. Combining this with the dimensions of acceleration gives the dimensions of force as ML/T^2. To come to a complete stop in one second, the individual in the example in the previous paragraph would have had to exert a force of just over 100 kilogram meters per second each second. The unit kilogram meter per second each second is the unit of force in the MKS system. It arises so often in science it has been given the name "newton."

The next concept to be considered is energy. One type of energy most people are familiar with is kinetic energy. This is the product of mass and velocity squared, divided by 2 or $E=MV^2/2$. By now, readers should figure out the dimensions of energy as ML^2/T^2 or MLA. In the example above, the 65 kilogram person, walking at 1.6 meters per second, has a kinetic energy of about 83 kilogram meters squared per second each second. One kilogram meter squared per second each second is a "joule" of energy. The preview gave the equivalent of joules in calories. The astute reader will note that a joule is a newton meter, or, more generally, energy is force times the distance through which it acts.

We also will deal with the important concept of work. This is the energy in a specific task. For example, if a kilogram falls one meter in the Earth's gravity field, then nearly 10 joules of energy are available to do some useful work, as the gravitational acceleration at sea level in the MKS system is 9.8 meters per second each second. However, by Mother Nature's Second Law, the amount of energy used to perform the task is always greater than the energy of the task. She won't let you extract 10 joules of work from the falling mass.

Finally, we come to power. This is the rate that energy is expended over time. The MKS name for power is watt, which is a joule per second, or dimensionally ML^2/T^3. Consider the example above of the 65 kilogram individual walking at 1.6 meters per second. If this individual comes to a complete stop in one second, then this person loses almost 83 watts of power. If she/he stops abruptly, say by hitting a brick wall, the loss of power would be much greater.

Table 2.1 summarizes the discussion in this section.

Item	Basic Dimensions	MKS Name
Distance	Length	Meter
Velocity	Length/Time	Meter/second
Momentum	Mass *x* Length/Time	Kilogram meter/second
Acceleration	Length/Time^2	Meter/second^2
Force	Mass *x* Length/Time^2	kilogram meter/second^2 or newton
Energy (Work)	Mass *x* Length^2/Time^2	kilogram meter^2/second^2 or joule
Power	Mass *x* Length^2/Time^3	kilogram meter^2/second^3 or watt

Table 2.1 Synopsis of MKS System.

Types of Energy

Although scientists are quite comfortable with the notion of energy, different disciplines have developed specialized energy vocabularies. Moreover, individual scientists are not always consistent in usage. The types of energy discussed below are a composite taken from Feynman's lecture at Cornell in 1964, "The Great Conservation Principles," and the index of *The Feynman Lectures On Physics*. As you will see, there is some redundancy in this list.

CHEMICAL. This is the energy associated with the binding of different atoms together to form molecules. This energy can be released during chemical reactions such as the burning of fuel. During this process, the chemical reactions produce new molecular structures; however, the number of atoms participating in the reactions stays the same.

ELASTIC. This is the energy that is stored in the molecular structure of substances. When the substance is stressed, such as when a dangling spring is loaded with a weight, the molecular structure warps to accommodate the additional force. When the weight is removed, the substance returns to its original state. Common manifestations of elastic energy are earthquake waves and sound. Elastic energy differs from chemical energy in that the molecular makeup of the substance is unaltered. In many college-level basic physics texts, elastic energy is sometimes referred to as elastic potential energy.

ELECTROMAGNETIC. This is the energy stored in electric and magnetic fields. At an elementary level this energy is associated with the position and flow of electrons. Harnessing this energy to do work is the basis of the telecommunications, computer, and electric utility industries. In some basic physics texts this is also classified as a form of potential energy.

HEAT. This is the energy commonly associated with the temperature of a substance. If the substance is heated and its temperature rises, then work can be performed as the substance cools. However, modern statistical mechanics has taught us that the temperature is simply a convenient macro characteristic of the kinetic and potential energies of the molecules that make up all substances.

INTERNAL. This is a catchall term usually used to describe the energy contained in the internal structure of substances that make up any particular system. This includes the various forms of energies of subsystems and constituents. This usually is the energy that is converted to work. Heat was the first type of energy identified as internal energy.

The relationship between heat or internal energy and temperature can be confusing, since it varies widely from substance to substance. Water, for example, has one of the highest heat capacities of naturally occurring substances. This means that when warmed, it can store a lot more heat energy than most other substances, and yet it does not show as strong an increase in temperature.

KINETIC. This is the energy associated with the motion of a substance. It is calculated from the formula given in the previous section.

MECHANICAL. This form of energy is associated with work performed by mechanisms for specific tasks, such as the energy delivered by pistons to turn driveshafts that turn the wheels that propel cars. J. P. Joule made one of the most remarkable discoveries of the 19th century. After tedious experiments, he showed that work, regardless of its form, could be converted into heat. This gave rise to the notion of the mechanical equivalent of heat, or, as I prefer to call it, the work equivalent of heat.

Much of industrial output involves transformations of chemical, internal (heat), or electromagnetic energy to kinetic (mechanical) energy to work. Most of the chemical energy now comes from fossil fuels. Prior to the Industrial Revolution, the energy transformation sequence was chemical to kinetic to work with the chemical energy coming from workers and beasts.

NUCLEAR. This is the energy stored in the elementary particles that make the nuclei of atoms. This energy is manifested in radioactive decay of materials. It is associated with the "weak" force in quantum mechanics.

POTENTIAL. This is the energy a substance acquires from its position in some force field. The most common example is the energy required to lift an object some distance in the Earth's gravity field. As mentioned earlier, one could also view the displacement of electrons away from the nucleus of an atom as a form of potential energy. Changes in potential energy are the important factors in a system's energy balance. If a body increases its distance from the center of the Earth, its potential energy is increased. The potential energy associated with raising and lowering of objects in the Earth's gravity field was one of the first forms of energy harnessed by our ancestors.

RADIANT. Along with relativistic energy, this is perhaps the most basic form of energy. As noted in Chapter 1, radiant energy is carried at the speed of light in discrete packets of pure energy called photons. The amount of energy in a packet is inversely proportional to its wavelength.

RELATIVISTIC. This is the ultimate amount of energy that may be extracted from a substance. It is given by Einstein's famous equivalency between energy and mass, $E=MC^2$.

As suggested at the beginning of this section, this categorization is somewhat arbitrary. What is not arbitrary is the First Law: Energy can only be transformed from one type to another type; it cannot be created or destroyed. In essence, all of these energies are equivalent. That is, a joule of energy is just that; it doesn't matter if it is kinetic, potential, relativistic, or any of the other types listed above.

Absent from this list is the specific mention of work as a form of energy. In high school science courses work is often taken as equivalent to mechanical energy. Keep in mind that, here, work will be used in a broader context than just mechanical energy. But no matter the context, the First Law assures equivalence in terms of any of the other forms.

Mother Nature's First Law

Everyday experience shows that the energy balance for most systems is not constant. Don't be confused by this; there is no violation of the First Law. When one applies the First Law to systems, it is necessary to account for the exchange of energy between the system and its environment. This was discussed in Chapter 1 and shown schematically in figure 1.1.

In order to explore this further, appropriate mathematical statements of the First and Second Laws for a system are required. In arriving at these statements, I have relied on concepts from classical irreversible thermodynamics (called "thermodynamics" by some). This approach focuses on a system's entropy and energy changes in time. These changes arise from flows across the system's boundary and from the transformation of various forms of energy into internal energy and the production of entropy within the system. For most chemists and many other scientists as well, the term "thermodynamics" is associated with equilibrium systems and reversible processes, and extension to irreversible paths implicit in the approach followed here is not strictly justified. Chemists also are often interested in systems evolving among states of greatly differing energies such as in most chemical reactions. For the systems of interest here, and for purposes of this book, the explicit introduction of time, as well as energy and entropy fluxes, are pedagogically advantageous. Expert readers may argue the details, but will agree that predicted evolutions of natural systems on Earth always are accompanied by the generation of entropy within the system. Proposals that suggest otherwise are highly suspect at the very least. If such claims are ever made, it is especially imperative that the scientific protocol be followed to determine all sources of this inconsistency.

With this caveat in mind, an appropriate statement of the First Law is:

$$\frac{\partial E}{\partial t} = T_E + G_E.$$

The left-hand side of this equation is the rate of change of the internal energy E of the system with time *t*. Remember that the rate of change of energy is just power. In many applications, E might include chemical, elastic, and heat energies perhaps lumped into an "internal" energy. The first term on the right-hand side of the equation, T_E, accounts for the energy supplied (or lost) to the system by transport across the system boundary. This includes advection, conduction, radiation of heat, and energy transfer arising from mechanical interactions between the system and the environment. For example, if the system is a bicycle pump, then the energy expended to push the handle in adds to the internal energy of the air in the pump. Of course, a system can transfer energy to the environment. One example is the expansion of gas in the cylinder of an engine to move the piston. Note that if the system is isolated, then T_E is exactly equal to zero. For closed systems this term includes only electromagnetic, potential, and radiant energy. Clearly, for open systems, all forms of energy are included.

The other term on the right-hand side, G_E, accounts for the gain or loss of internal energy by energy transformation processes acting throughout the system instead of across the boundary. Examples are energy released (or used) by chemical reactions within the system and the conversion of kinetic energy to heat through friction or viscous processes. This term could also include nuclear or relativistic energies if such processes are active. G_E is often referred to as a "source" term for internal energy. From the system energy balance standpoint, it accounts for the transformation of energy from one type to another.

The two terms on the right-hand side can be positive, negative, or zero. Of course, positive values indicate an increase in time of internal energy, while negative values indicate a decrease. If the sum of the two terms is zero, then the time rate of change of internal energy for the system is also zero.

Now let us examine a special case of the conservation of energy equation. Suppose the system is not in motion, hence it possesses no kinetic energy or exhibits no change in potential energy. Furthermore, neglect chemical, elastic, electromagnetic, nuclear, and relativistic processes. Then the processes that can alter the energy of the system are heat and mechanical energy inputs across the system boundaries. In the 19th century this special case was the subject of many experiments. These experiments led to the realization that there was a mechanical equivalent of heat. Later this was generalized to the concept of a quantity, energy, which was conserved.

Energy Transformations

We often perform convoluted energy transformations to provide zest to life. Consider the transformations involved in bungee cord jumping. Bungee cords are super-elastic ropes made from special polymers. They are capable of stretching to several times their original length when stressed. Because of this property, they are, in some circles, popular vehicles for recreation. Picture yourself on the New River Gorge Bridge in West Virginia overlooking the New River Gorge, hundreds of feet below. Tie a bungee cord to the bridge, then to yourself, and then leap off the bridge. As you fall, you are losing potential energy at an alarming rate. At first the potential energy is converted to kinetic energy as manifested in the speed at which you are falling. However, after you have fallen the length of the cord, this kinetic energy, as well as further loss of potential energy as you continue to fall, is converted to elastic energy manifested in elongation of the cord. Thus, your rate of fall is slowed, and ultimately you stop when the potential energy difference between the bridge and the stopping point has all been converted to elastic energy in the cord. Then the elastic energy is converted to kinetic energy as you start an upward rebound. The rebound will almost take you back up to your jumping-off point. Some of your kinetic energy is transferred to increased kinetic energy of adjacent air molecules as you fall and rebound and modest heating of the cord as it is stretched. Thus, you won't exactly return to the original elevation; unless you plan the jump so that at the full extension of the cord your feet touch a rock in the gorge several hundred feet below. By kicking off the rock, you provide an extra increment of kinetic energy to make up for the energy that is used to overcome air resistance. So on the rebound; you could step back onto the bridge. If your kick off is too strong, you may even end up above your jumping off point. Despite the adrenaline rush and all the energy transformations, no energy is lost or created when you return to your jumping off spot.

The conservation of energy is such a powerful principle in science that it is frequently used as the basis of tests for Ph.D. candidates. During these exams, the candidates are asked thought-provoking questions related to their disciplines. These usually take the form of conceptually simple, but often impossibly difficult, experiments to perform. Generally, the intention is not to encourage the experiments; however, I know of several cases where answers led to refereed publications. Rather, they are designed to test the candidate's knowledge of basic assumptions underlying appropriate theories, and the ability to apply critical thinking in answering the questions. The practice of asking students these types of questions goes back to at least the time of Socrates.

Albert Einstein used this approach in a long-running dialogue on quantum mechanics with Niels Bohr. Here he used the expression "gedanken" (German for "thought") experiments. It is ironic that gedanken experiments are associated with Einstein when he lost all the arguments with Bohr.

Here are two of my favorite problems. One reason for my partiality is that I correctly answered them when they were posed by my advisor, Professor Robert O. Reid, on my Ph.D. exam.

Question 1a: Take a railroad boxcar suspension spring and put it in an insulated chamber. Now compress the spring. What happens to the energy used to compress the spring?

Answer: That energy appears as an increase in the strain energy of the spring. Essentially, the energy used to compress the spring deforms its molecular structure.

Question 1b: Now suppose you pour into the chamber a strong acid that completely dissolves the spring. What happens to the strain energy increment?

Answer: This will appear as an incremental increase to the internal energy of the solution. Presumably the temperature would be higher in this case than in the case where an uncompressed spring was dissolved. However, the increase in temperature may be masked by heat released or consumed in the chemical reaction of the acid and steel.

I recently posed this question to a larger audience. Wayne Schroll pointed out that photons or light could be released as well in this case.

Question 2: Take two insulated beakers of fluid at different temperatures. Put the warmer beaker over the cooler beaker. Now allow the two fluids to mix, that is, come to thermal equilibrium. The center of mass of the unmixed initial state is lower than the center of mass of the equilibrium state. Where did the energy come from to raise the center of mass of the mixture?

Answer: The increase of the potential energy of the mixture has to come from a decrease in the internal energies of the original two fluids. That is, the temperature of the mixture in this experiment would be less than the temperature of the mixture obtained by placing the beakers side by side and allowing them to mix.

Of course these problems are vastly oversimplified as they neglect a lot of factors that would be important if the experiments were actually performed. In reality, it would be impossible to devise containers that wouldn't exchange some heat with the constituents while they were evolving to a new equilibrium. Moreover, since both problems involve mixing, a complete analysis would have to account for a temperature increase arising solely from the mixing.

Complications in seemingly mundane problems are a characteristic of science. No matter how simple one tries to make a problem, it is always possible to drill down to more basic issues. The practice of science is like peeling the skin off an onion: answering one set of questions often reveals a whole new set.

Aside on Orders of Magnitude

The section following this is the first of a number of sections where we will be comparing numbers that vary greatly. This is an issue that has arisen so often in science that a special vocabulary has evolved. Since this vocabulary is mystifying to many nonscientists, it is appropriate to explain it. The Rosetta Stone that simplifies this is powers of 10. This is just like our currency system. Consider first a penny and a dollar. Since one dollar is one hundred pennies, and since one hundred is ten times ten, or ten squared, then a nerdy scientist would say that a dollar is two orders of magnitude more money than a penny. Similarly, a million dollars is six orders of magnitude more money than one dollar, and eight orders of magnitude more than one cent.

It generally is convenient to compress a large or small number by writing it as a number between one to nine times the appropriate power of ten. Thus, instead of writing \$2,000,000 I might be tempted to write \$2 x 10^6. The superscript on the 10 indicates the number of zeros to add after 2.

So much for numbers bigger than one. For numbers less than one, the superscript will be negative. This denotes the number of zeros after the decimal point. To illustrate this, compare my credit card limit to that of Bill Gates. Mine is \$10,000, or $\$10^4$. I don't know Mr. Gates's limit, but \$100,000,000, or $\$10^8$, seems reasonable to me. In case you get tired of counting zeros this is one hundred million. My limit, when divided by his limit, is 10^{-4}. In other words, his credit card limit is at least four orders of magnitude larger than mine.

Keep in mind that orders of magnitude are completely neutral. They are useful when comparing the relative sizes of numbers. Attaching significance or meaning to the relative sizes is a separate issue.

Energy Gauges for One Kilogram

The astute reader may have noticed a hierarchy of energy types associated with different scales of matter. At the bulk or largest scales, potential and kinetic energies are shared nearly equally by all parts of a substance. These are directly attributed to the substance's position and motion. Elastic energy is associated with the deformation of the substance, and ultimately is connected to the deformation of its molecular bonding. Heat energy is manifested in the motion of the molecules that make up the substance.

In the energy categories mentioned in the previous paragraph, the substance retains its chemical structure when the various energies are released to perform work. This is not true for energies associated with smaller scales of matter. Chemical energy is tied up in the molecular bonds of the substance. When this is released the substance changes its chemical nature. The atomic structure, however, is unchanged. Electromagnetic energy is associated with the electrons that comprise the atoms that make up the molecules of the substance. Finally, we come to radiant and relativistic energies. The former is comprised of the most elementary packets of energy possible, while the later deals with the ultimate equivalency of mass and energy.

Generally, processes operating on smaller scales of matter involve more energy per unit mass than do those operating on larger scales. For example, converting all the mass in a kilogram of wood to energy releases more than a billion times more energy than merely burning it. Table 2.2 provides a gauge for the amount of energy available from various processes in one kilogram of matter.

Energy Type	Example	Amount of Energy (joules)
Potential (Gravity)	1 M Displacement	10
Chemical	Wood	2×10^7
Chemical	High Grade Coal	3.5×10^7
Chemical	Typical Crude	4×10^7
Electromagnetic	Electrons	1.7×10^{11}
Nuclear	^{238}U	9×10^{13}
Relativistic	Total Annihilation	9×10^{16}
Chemical	Ben & Jerry's Chocolate Chip Cookie Dough Ice Cream	1.5×10^7

Table 2.2 Energy Gauges for One Kilogram

Perspective on Energy

Table 2.2 summarizes the efficiency of the technology for converting energy to work in terms of a unit amount of "fuel." Consider Neolithic societies in existence approximately 10,000 years ago. The technology for converting energy to work available to individuals, at that time, came from the conversion of potential energy to kinetic energy from falling objects; or it came from the kinetic energy generated by the individuals, beasts of burden, and perhaps fire. Of course, a variety of tools were used to increase efficiency. Still, the amount of energy an individual could use to perform work was in the range of the amount of energy he/she or beasts of burden could consume in eating. For humans, this is of the order of 10^7 joules per day and much of this is used for metabolic functions.

Thus, a convenient metric for conversion of energy to work for Neolithic society is a kilogram mass falling one meter. As seen by the table, this makes available about 10 joules. If we take a time scale for the fall, and conversion process of one second, then each operation releases no more than 10 watts of power. Individuals would be limited to a few hundred such operations per day. In essence, a Neolithic peon working as hard as he/she was able could only access as much energy in one day as we can obtain now by burning a kilogram lump of coal in a few minutes.

This was pretty much the situation for humanity until the start of the 19th century. Until then, individuals did not have the technology to utilize much more energy than did our Neolithic ancestors. Of course, tools such as levers, pulleys, etc. were very important in increasing individual efficiency, as were beasts of burden. The essential point, however, is that up until the Industrial Revolution, humans could not perform much more work than individuals in Neolithic societies.

This changed rather abruptly on historical time scales when steam was harnessed to do work. Steam engines provided substantially more energy than could humans or beasts. The introduction of steam technology resulted in widespread commercial applications, with consequent increased demands for more efficient energy sources. At first, wood was the fuel of choice. However, the huge increase in demand for fuel quickly produced a demand for more efficient fuels. This spurred scientists and engineers to learn how to tap into the chemical energy of coal. As shown in Table 2.2, coal produced more energy per kilogram than wood. Of even greater importance, coal allowed significantly higher temperatures to be achieved in furnaces. Higher temperatures not only produced more efficient engines, they also revolutionized ironmaking. The shift to coal as the primary fuel for steam engines also marked a significant change in the geopolitical balance of power. Countries such as Great Britain, the USA, and Germany, with vast reserves of coal, gained enormous economic and military advantages.

As has been well documented by historians, the ascendancy of industry during this period was a catalyst for social change as well. Curiously, Sardi Carnot, the first person to carry out definitive experiments on energy and entropy, foresaw this. Less than 90 years after his death in 1832, democracies replaced centuries-old monarchy rule in all of the industrialized countries. His and other scientists' efforts to quantify the First and Second Laws of Mother Nature provided the basis for designing ever more efficient steam engines. This, in turn, was the catalyst for social change.

Another interesting synergism developed during this period. The ascendancy of steam power and consequent mechanical efficiencies revolutionized transportation. Not only did transportation costs plummet, but the speed of transportation and the amount of cargo that could be carried increased dramatically. Before the start of the Industrial Revolution the distance one could travel in one day was measured by how far one could travel by horse or stage. Typically, this was of the order of 20 miles, which was not much different than the distance Neolithic societies could travel. Within 50 years, railroads had stretched this distance to hundreds of miles. Furthermore, trains could transport vastly more cargo than animal-drawn vehicles.

This heralded a major shift in economics. It became more efficient to mine and transport fuel to where it was needed rather than forage for indigenous fuels as was the common practice before the Industrial Revolution. At the start of the Industrial Revolution, wood was the only fuel available. But as technology evolved, it became possible to use coal, which was more efficient and could achieve higher temperatures in the furnaces than wood. By 1875, coal surpassed wood as the dominant energy source in the industrialized nations. From Table 2.2, the unit amount of energy one could tap into increased by a factor of 10^6, or a million times more than could be accessed only 75 years earlier. In economic terms, this technical achievement meant that the price people paid for energy dropped by a corresponding amount. Energy, and hence work, became cheap.

The search for even more efficient and plentiful fuels continued. The reign of coal lasted 90 years. By 1965 oil and natural gas replaced coal as the dominant source of energy in industrial societies. It is important to note that the shift to petroleum was not because coal resources were used up. Indeed, coal is still our most plentiful fuel. The shift to petroleum occurred because it delivered more energy per unit mass than coal, it was easier to transport, and it was better at providing energy, and hence work, on demand. Cold starts of steam engines take time, whereas cold starts of gasoline engines are virtually instantaneous.

There is another advantage of petroleum-based fuel over wood or coal. It is significantly less polluting! Petroleum fueled engines are much smaller than their steam counterparts, hence less energy is lost through heat. This means less fuel is consumed, and the exhaust has far fewer particulates that get in the air. Up through World War II, London was well known for smog and poor air quality. The widespread shift to petroleum greatly improved air quality there and in other large cities of the industrialized world.

About 50 years ago, scientists and engineers were able to establish another million-fold increase in the amount of energy released by a unit amount of fuel. This, of course, was the harnessing of nuclear power. Existing technology is capable of converting approximately 0.1% of the mass of the fuel directly into energy. Thus, we are only three orders of magnitude away from the theoretical limit for the amount of energy available from a unit amount of fuel established by Einstein's equation. As this is the most energy that can be extracted from a unit amount of fuel, there will be no more million-fold increases in the amount of energy we can obtain from matter.

The rise of energy consumption and associated technology with the industrial age also marked a shift in public attitudes. Up until the start of the Industrial Revolution, most societies had a healthy respect for Mother Nature. There was deep-seated fear of natural processes and the awesome power of nature. In this regard, the attitudes in the pre-Industrial Revolution period may not have been all that different from those of Neolithic societies. But once the Industrial Revolution was underway, an attitude that science and technology could "conquer" or "master" nature arose. Moreover, as this myth goes, science and technology put humankind above natural laws. With new materials we could build with impunity in any location no matter what the natural hazards were, and science and technology could handle any deleterious consequences of the enormous uses of energy. This view is wrong. In reality, scientists and engineers merely learned how to use the laws of nature to accomplish desirable goals. Humankind cannot tame nature, but with hard work and luck, we can learn how to use natural processes to our advantage.

During most of history, humankind has traveled to energy sources. Paleolithic peoples hunted for food, and medieval societies depended on the efforts of serfs, animals, and running water for energy. If the game disappeared, the forests were cut down, or the streams dried up, then people had to migrate to other favorable locations or suffer substantial decreases in their standards of living.

This changed with the Industrial Revolution. Instead of communal societies being more or less self-sufficient in the food they grew and the products they manufactured and sold, it became cost effective to concentrate agriculture and manufacturing capabilities in the most favorable locations and rely on cheap transportation to get products to market.

The Industrial Revolution was the real start of economic globalization. The enormous increase in the energy available from a unit amount of matter, and the consequent increased sophistication of the systems used to convert energy to work, led to a new energy policy. Starting in the 1850s it became more efficient to extract the energy ores where found and transport them to the consumer. Today, there are electrical power and natural gas lines crisscrossing the nation bringing vastly more energy directly into our homes than was conceivable 100 years ago. There are supertankers and large colliers transporting hydrocarbons all over the world. Nevertheless, access to energy is still a dominant cause of conflict. Unfortunately, as the amount of energy available from fuel has increased, so has the level of conflict. Like it or not, this is the way society operates.

Esoteric energy technology is so commonplace in modern life we tend to take it all for granted. Nevertheless, a lot of intricate energy transformations are involved in even the most mundane tasks. Take, for example, driving a car to the grocery store. The work to be accomplished is to mobilize the car to transport you to the store. To do this, you first turn on the ignition system, which converts electrical energy stored in a battery to mechanical energy that turns the engine over so that it can start. The engine then converts chemical energy stored in the gasoline to mechanical energy that drives the pistons that turn the crankshaft that turns the differential that spins the wheels that propel the car. Depending upon driving conditions, perhaps 10% of the chemical energy of the gasoline ends up in the kinetic energy of the car as it moves to the grocery store. The rest of the energy is used to overcome friction as the pistons move in the cylinders; friction in the various mechanical linkages; friction between the tires and the road; air resistance on the car; incomplete combustion of the gasoline; energy to vent the combustion products; energy transmitted back to the environment in the form of photons; heat conduction; and energy to run the car accessories such as air conditioning, the CD, the cooling system, lights, various engine and accessory monitoring systems; and finally, the energy to recharge the battery.

Now consider what happens when you stop the car at the grocery store. The car has considerable kinetic energy that must be converted to some other form of energy if the car is to be stopped quickly. This is the role of the brakes. They act to convert the kinetic energy of motion to energy in the form of heat.

Note that during all this time the car is expelling mass through the exhaust to its environment, that is, it is an open system. One of the exhaust products is carbon dioxide (CO_2).

Cheap energy altered community economics and opened up other economic opportunities. By focusing energy to alter natural materials to produce substances with desirable special properties, metallurgy and material sciences were revolutionized. This development allowed the construction of large and efficient engines, machines, and structures. In turn, this produced more synergism.

Up to now the discussion has focused on the tremendous increase in energy the Industrial Revolution has made available to us. But there is another aspect to this story. This is the enormous increase in the amount of power available. Neolithic peoples might have been able to transform 10^7 joules of energy to work in one day. We can do that much in a few minutes by burning a kilogram of coal. So not only do we have access to more energy than people did a couple of centuries ago, we also have the technology to transform enormous amounts of energy to work on demand in short periods of time. The increase in energy usage has been paralleled by a comparable increase in availability of power. Compared to civilization 200 years ago, individuals today not only have access to orders of magnitude more energy, but they can convert it to work orders of magnitude faster.

Suggested Reading

R. P. Feynman, R. B. Leighton, and M. Sandy, *The Feynman Lectures on Physics*, Addison-Wesley, 1975. This is one of the most highly regarded textbooks in science. It is based on an underclass physics course taught by Richard Feynman at Caltech. The book represents a substantial team effort by senior faculty and graduate students from several disciplines. It is the only introductory text I am aware of that is mandatory reading for physics graduate students preparing for candidacy exams. The discussions of basic physical processes are quite lucid, so I recommend it to nonscientists as well.

R. P. Feynman, *The Character of Physical Law*, The M. I. T. Press, 1965. This is the edited version of a series of public lectures Feynman gave at Cornell. Most physical scientists are familiar with this material; however, it is written at a level that will appeal to nonscientists.

M. Y. Han, *Quarks and Gluons*, World Scientific, 1999. Those interested in more detail about the powers of 10 should consult this book. Appendix 2 gives a very readable explanation.

3. The Faces of Entropy

"The law that entropy always increases holds, I think, the supreme position among the laws of nature. If someone points out to you that your pet theory of the universe is in disagreement with Maxwell's equations—then so much the worse for Maxwell's equations . . . but if your theory is found to be against the second law of thermodynamics, I can give you no hope; there is nothing for it but to collapse in deepest humiliation."

A. Eddington

Preview

The Second Law quantifies many practical and philosophical issues that have always been part of the human cultural heritage. It is the reason why there are no perpetual motion machines, there are sanitation systems, we age, there is industrial pollution, we cannot unstir coffee to recover the sugar and cream, and why there are residual products when cigarettes are smoked. It touches on deep philosophical questions like irreversible actions; why we remember the past and dream of the future; and even cause and effect. Many ancient fables, such as Pandora opening her box, convey the gist of the Second Law. Entropy ranks up there with love, hate, and honor as one of the most profound concepts humankind has ever devised. One goal of this chapter is to illustrate the scope of the Second Law in our culture.

Entropy is derived from a Greek word, meaning "turning" or "transformation," according to the *Oxford English Dictionary*, 2nd edition. Rudolf Clausius, one of the leading scientists of the 19th century, who played a major role in quantifying the Second Law, first used it in modern times. Unlike energy, the *OED* gives only two definitions of entropy, which are the scientific definitions. Consequently, this chapter is organized around these definitions. The second definition, which is taken up first, involves the notions of order and disorder. Statistical mechanics and information theory are based on this definition. Many applications of this definition are used in economics, sociology, art, philosophy, and psychology. These applications constitute the probabilistic faces of entropy. The first *OED* definition is the basis for the classic faces of entropy. Among other things, it identifies entropy as a measure of the unavailability of all the energy of a system to do work. Both definitions were developed in the 19th century; however, the use of entropy as a measure of the unavailability of energy to do work predates the statistical definition. It is emphasized here and later that the two definitions are equivalent; they are merely two different mathematical and conceptual approaches to the description of a most fundamental principle of nature.

Although the two *OED* definitions are more closely connected with science and technology than are the definitions of energy, neither indicates the depth to which the concept of entropy is now rooted in our culture. An Internet search of "entropy" should convince the reader of this. There are thousands of web sites that connect entropy to topics ranging from traditional applications in science and engineering, to the disciplines mentioned above, and even the occult. The search reveals, for example, the Society for Studies on Entropy. This society is composed of academics from Japan interested in applications of entropy outside of engineering. At a personal level, I have met a number of nonscientists who profess ignorance of science, yet staunchly claim expertise in aspects of entropy such as order and disorder.

The unique role of the Second Law in science means that entropy can be a contentious topic among scientists. Many scientists (and nonscientists) believe there is only one way to determine entropy, which just happens to be the one they use. My view is that entropy, like energy, is too basic to have a single physical interpretation. So, in order to avoid unnecessary pedagogy, I will follow the program of Harold Grad, who noted that it is not always possible to assign a unique value of entropy to a system (see citation at the end of this chapter). There may be many different ways to assign values of entropy and all of them may be worthwhile. Grad asserts:

"The proper choice will depend on the interests of the individual, the particular phenomena under study, the degree of precision available or arbitrarily decided upon, or the method of description which is employed; and each of these criteria is largely subject to the discretion of the individual."

The freedom to choose an appropriate entropy might be regarded as the "first amendment" to the Second Law. The unifying concept for different entropies is irreversibility. Isolated systems evolve from states of low to high entropy regardless of how one elects to determine entropy. A system at its maximum entropy is in some sort of equilibrium condition and cannot evolve further without an input of energy and/or export of entropy.

Readers may have detected a subtle, but fundamental, difference in the philosophical views of the faces of energy and entropy taken here. The faces of energy; chemical, heat, kinetic, etc., are unified by the requirement that they are conserved in any transformation. A joule of heat energy is the same as a joule of kinetic energy. This indistinguishability is the essence of the First Law. The faces of entropy are unified by the requirement of irreversibility. The essence of the Second Law for isolated systems is that the energy transformations are irreversible.

Here I will be selective and restrict attention to just applications in science and technology. As stated earlier, we follow the two *OED* definitions and categorize entropy by concept and mathematical technique rather than by application. The probabilistic faces comprise the most popular category if the number of web sites is used as a metric. This is taken up next. Then I discuss traditional faces of entropy. After this is an aside on Carnot engines. The stage is then set for a remarkable result devised by Ilya Prigogine. We will also appeal to this theorem in Chapter 2 of Part 2. Then I discuss the role of entropy in change, followed by asides on the arrow of time, causality, and irreversibility. The chapter concludes with a review and debunking of attempts to "prove" violations of the Second Law.

The Probabilistic Faces

Macrostates, microstates, and distinguishability

In order to appreciate the significance and broad appeal of this approach, it is necessary to introduce some more terminology. First, recall from Chapter 1 the concept of systems and subsystems. The notion of systems and subsystems applies nicely to the traditional faces of entropy where the assumptions are that the number of subsystems is small and observable, at least in principle. With the probabilistic faces, neither of these assumptions is true. Instead, the language used with the probabilistic faces refers to macrostates and microstates. For nonscientists, macrostates and microstates correspond loosely to the states of systems and subsystems, respectively. Keep in mind, however, the concept of macrostates and microstates as now used has different applications and implications than systems and subsystems. Sometimes it is useful to think of macrostates as observable conditions and microstates as unobservable.

Let us illustrate this by starting with microstates. A regular die used in parlor games and gaming tables has six faces, each with one to six dots. Think of each of these six faces as a microstate of the die. Any throw will show one of these microstates as an observed macrostate. If the die is fair, then any one of the microstates is equally likely to occur when the die is thrown.

The more interesting and practical situation is when several different microstates can produce the same observable or macrostate. This is illustrated by throwing two dice simultaneously. The macrostates are the total number of dots shown in the throw. There are eleven possible macrostates or outcomes, the numbers 2 through 12. However, as readers either know or can readily verify, there are 36 possible microstates. The macrostate 2 can only be produced if both dice show 1. Similarly, macrostate 12 occurs only when the dice show 6. All other macrostates are produced by more than one combination of the dice. For example, the macrostate 3 can occur when die A shows 1 and die B shows 2 or vice versa. So even though the dice are fair, that is, each of their microstates are equally probable, when both are thrown simultaneously, the outcomes or macrostates are not all equally probable. As most readers already know, macrostate 7 is the most probable outcome since it can be achieved by the most number of microstates (6).

If one is gambling on the outcome of a throw the microstates are superfluous; the only issue is the odds of achieving a desired (or undesired) macrostate. More generally, knowledge of a macrostate generally does not provide much information on the specific microstate from which it was produced.

Now consider a system with a much larger number of microstates, a typical PC. Each bit of information stored in the computer can be thought of as a microstate. Denote the total number of microstates, or bits of information, by the letter m. Typically m is at least 10^8, or 100,000,000. Since the computer is digital, there are approximately 2^m microstates. This is 2 raised to the hundred millionth power, a really, really big number. When the computer is turned on, the microstates are activated to produce the starting view of the screen. Think of the screen image as a macrostate. In its lifetime, a computer may be turned on or otherwise activated thousands of times. The initial screen image or macrostate is always the same, but the microstates that produce the same macrostate are never the same. The times when the screen is activated can never be the same, the saved files generally are not the same, and a lot of invisible housekeeping programs are always running in the background. These and many other factors alter the microstates of the computer, yet the initial screen image is always the same.

After activating the computer, suppose a word processing program is launched. The screen image is a blank tablet with the cursor blinking while waiting for the next instruction. Suppose further that the user enters "entropy" and then deletes the word. The screen image or macrostate is exactly the same as it was before "entropy" was entered, but the two corresponding microstates are not. Not only are the processes mentioned above constantly altering the microstates, "entropy" has been moved from one part of the computer memory to another part. In fact, the microstates are ever-changing, even if the operator does nothing and the macrostate stays the same!

There are a couple of messages in these two examples. First, it is useful to think of a macrostate as a characterization of a large number of microstates. Changes in the microstates do not necessarily translate into changes in the macrostates. In other words, one macrostate can be produced by more than one microstate. But each microstate can only be associated with only one macrostate.

The other message is distinguishability. In principle, all macrostates are observable or distinguishable; that is, each can be assigned a unique name. Microstates may be either distinguishable or indistinguishable. The 36 microstates of the outcome of the throw of the dice are easy to distinguish if one makes just a modest effort. Many of the microstates of a PC are distinguishable, but naming many others, like the exact location of "entropy" after it was deleted, can be determined only with expert knowledge and after considerable effort. The positions and velocities of individual molecules in a gas are one of the classic examples of indistinguishable microstates. There is no experimental means to measure such properties of individual molecules in the gas. Without information on its location and velocity, one molecule of the gas is exactly the same as any other.

Even if such measurements were possible, the amount of data is overwhelming. To illustrate this, suppose one makes a gas out of a mass of material, whose weight in grams (a thousandth of a kilogram) numerically is the same as the molecular weight of the material. Furthermore, the temperature of the gas is set at the freezing point of water (0° Centigrade) and the pressure at the atmospheric pressure. The gas will then occupy 22.4 liters (about 137 cubic inches). The observation of the position and velocity at one time of each one of the molecules requires over $5.4 x 10^{24}$ entries. To be meaningful, the data should be sampled at least at a millionth of a second intervals. One minute of data thus provides over $3 x 10^{33}$ entries. This is about 10 orders of magnitude larger than the number of visible stars.

Incidentally, the mass of material whose weight in grams is numerically equal to the molecular weight of the material is called a gram-mole. For example, the atomic weight of carbon, C, is 12, so a mole of pure carbon weighs 12 grams. Nitrogen, N, has an atomic weight of 14, so a mole of the molecule N_2 weighs 28 grams. Most nonscientists are surprised that the volume (at atmospheric pressure and 0/ Centigrade) and the number of molecules in a gram-mole of gas are the same for all gases. The number of molecules in a gram-mole is called Avagadro's number, and it is slightly more than $6 x 10^{23}$.

In many elementary applications the issue of distinguishability is clear-cut. If one can devise a rule to name the microstates, then they are distinguishable. Sometimes the application of the rule is so difficult, it may be judged impractical to implement. Often the results are not sensitive as to whether the microstates are distinguishable or not. In other cases, distinguishability can be a murky issue. For example, a box of identical objects is distinguishable if the precise locations of the objects can be specified. But if the objects are the molecules of a gas, and thus in a constant state of motion, they are indistinguishable. Realistic applications often have a mixture of distinguishable and indistinguishable microstates. Consequently, I have found it useful to adopt Grad's philosophy, cited earlier, on definitions of entropy. Whether or not the microstates are distinguishable depends upon the goals of the research, the phenomena under study, the degree of precision desired, and the method of description used in the analysis. However, every decision about distinguishability must be consistent with no decrease in the entropy of the system when it is isolated.

The concepts of macrostates, microstates, and distinguishability are endemic in virtually every aspect of modern technology. Perhaps readers will now be able to look at typical home or office appliances, and imagine the multitude of complexions required to make these devices "user-friendly," and ponder, as I regularly do, which are distinguishable.

Boltzmann's magic bullet

Development of modern technological devices requires engineers and scientists to quantify the relations between the microstates and macrostates. Ludwig Boltzmann, who lived in the latter half of the 19th century, was the first to do this. Although his motivation was to link macrostate thermodynamic properties of gases, such as temperature and pressure, to the behavior of the individual molecules, his ideas have been applied widely outside of thermodynamics and statistical mechanics. The magic recipe he developed is given by the deceptively simple formula:

$$S = k \cdot \ln W.$$

Here S is a measure of entropy of a particular macrostate; k is a universal constant (Boltzmann's constant) if the application is to thermodynamics; and W is a measure of the number of microstates that can produce the macrostate. The ln in this equation stands for the natural logarithm, which is the logarithm to the base of an irrational number that is slightly larger than 2.7. Another common logarithm base is 10; however, the distinction of bases is unimportant here. The important issue is that the logarithm function is another great compactor of large numbers, closely related to the powers of 10 described in Chapter 2. The logarithm to the base 10 of 1 is simply 0 while the logarithm of 1,000,000 is 6.

Boltzmann's equation provides the template for quantifying entropy from the knowledge of just the number or probability of occurrence of appropriate microstates. The macrostate produced by the largest number of microstates will have the largest entropy. Thus, the Second Law says that isolated systems will evolve to states of maximum entropy, that is, towards those states determined by the most microstates.

Each macrostate has a value of entropy that depends on the number of microstates that can produce it. In many applications in physics and chemistry all possible microstates have the same probability of occurring. Then W is simply the number of microstates capable of producing a particular macrostate. In general; however, specification of W depends on whether the microstates are distinguishable or not. Boltzmann assumed indistinguishable microstates in his research. On the other hand, modern quantum statistical mechanics presumes distinguishable microstates. One of the softballs Mother Nature has thrown us is that for many ensembles of large particles, the quantum and classical descriptions of W are nearly the same.

Although Boltzmann developed his formula as a model for the behavior of gases, the ideas are just as applicable to any situation where there is a need to quantify the macrostates of any system. His formula has much more practical utility and is the basis of many more web sites than Einstein's formula $E=MC^2$. $S=k \cdot \ln W$ is one of the great intellectual achievements of humankind.

Incidentally, Boltzmann's life illustrates the old saw that a person judged to be one or two years ahead of the times is described as brilliant, someone 10 years ahead of the times is a genius—but anything more makes the person a crackpot. Boltzmann came up with his recipe in the latter half of the 19th century, decades before the atomic theory of matter was widely accepted, and without the benefit of technology to provide experimental guidance! Although he held prestigious professorships and regularly published his research, he spent much of his time defending his work from vitriolic attacks by other scientists. These attacks contributed to periods of severe depression and likely were a factor in his suicide at the early age of 50. Boltzmann's equation is included along with his name and birth and death dates on his tombstone in Vienna. He is buried only a few meters from Beethoven, Brahms, Mozart, Schubert, and Johann Strauss. Many years after his death the deep significance of his work was finally appreciated. Einstein was more fortunate; his genius was recognized during his lifetime.

Boltzmann's cause probably was not helped by a paradox noted by Josiah Gibbs, a professor of chemistry at Yale. The paradox is illustrated by a gedanken or thought experiment. Take half a mole each of two gases of the same molecular weight and put them in separate containers. Carbon monoxide, CO, and the nitrogen molecule, N_2, are reasonable candidates since both have a molecular weight of 28. If the two containers are connected, the two gases will mix, and application of Boltzmann's equation shows that the entropy of the mixture is greater than the sum of the two constituents before mixing. This is called the entropy of mixing, a concept universally accepted by physical chemists. But if the CO container is replaced by another container of half a mole of N_2, the same formula predicts the same increase in entropy. This, of course, is ridiculous since there can be no diffusion or mixing in a homogeneous gas. It is ironic that Gibbs is credited with the paradox since he was one of the few scientists of the times who understood the deep significance of Boltzmann's research.

Many elementary textbooks on statistical mechanics claim the Gibbs paradox can only be resolved by quantum mechanics. In simple terms, this argument is that in classical statistical mechanics all gases with the same molecular weight are indistinguishable. In this theory, the chemical makeup of the gas has no effect on its statistical mechanical properties. On the other hand, quantum statistical mechanics looks more closely into the structure of the molecule. It considers the characteristics of the bonds holding the atoms of a molecule together. Thus, two molecules made up of two different chemical species will have distinguishing quantum properties. Simply put, the quantum theory says that it is impossible to transmute CO molecules to N_2. Quantum physicists and chemists rarely note that Grad resolved the Gibbs paradox with an alternate but equally plausible (yet distinguishable) W, without any appeal to quantum mechanics.

Since high entropy states imply large numbers of microstates whose details are largely unknown, it has become popular to equate entropy to "disorder." In this interpretation, the Second Law states that isolated systems evolve from states of order to disorder. For example, in the conversion of the chemical energy in fuel to work, an engine takes the orderly molecular arrangements of the fuel to produce work, and ejects fumes and other residues with considerably less order in their molecular structure.

This sounds nice, and it even has a certain poetic appeal, but one should not get caught up in vague notions of order versus disorder, nor should one equate lack of knowledge with disorder. One of the beauties of poetry is license, so keep in mind that there are no precise definitions of these terms in common usage. Thus, order and disorder may be in the mind of the observer. I shall try to illustrate some pitfalls.

First, we will do a gedanken experiment. Imagine a box with 1,000 slots, each of which is occupied by coins all turned to heads. Since there is only one microstate that can produce this macrostate, $W = 1$, and by Boltzmann's recipe the entropy S = 0. After the box is shaken vigorously we would expect that about half of the coins would be flipped to tails, the most probable state. With a little help from combinatorial arithmetic, one finds that this macrostate can be produced by 10^{300} microstates, that is, $W = 10^{300}$ and S = 300k, using logarithms to the base 10. The reader may imagine the appearance of heads and tails in a typical realization to be jumbled, and thus supportive of the notion of disorder. However, some of these 10^{300} microstates have a very orderly structure. These would include alternating sequences of heads and tails such as one head and one tail, two heads and two tails, and so on, up to 500 heads and 500 tails.

In real physical systems, large entropy states may be achieved by segregation of order to certain regions with large intervening regions of disorder. Such patterns may appear orderly. Computer models of isolated geophysical fluid dynamical systems, such as the atmosphere and ocean, evolve from initial states of many small vortices distributed randomly in space and with random intensities, to end states with just a few large vortices. The entropy of the latter state is larger than the entropy of the more disorderly-appearing state of many small vortices. As the small vortices merge to form large vortices, there are very intense viscous fluid interactions on small scales. These generate a significant amount of entropy. Of course, large/sparse vortex end states do not occur because the real ocean and atmosphere are open systems with substantial fluxes of energy and entropy at their boundaries. Certain chemical reactions display beautiful examples of structure or order. A few are given in Atkins' book cited in Chapter 1. The essential point is that real systems are not isolated and there is potential for flow of energy and entropy into and out of the system. The evolutionary path of any system is the consequence of a delicate balance between its starting state, its dynamical constraints, and the energy and entropy flows across its boundaries.

In our everyday lives, notions of order and disorder can be much simpler, and so the concept of evolving from a state of low to high disorder is useful. Consider, for example, mixing cream and sugar with coffee. Before mixing, each of the three constituents is isolated, so the system is in its most orderly and lowest entropy state. Stirring this mixture increases the disorder, thus increasing its entropy. Another noteworthy aspect of this experiment is that it will take a lot more energy to restore the mixture of coffee, sugar, and cream to the original unmixed state than it took to combine them.

Boltzmann's formalism has been used in communications to quantify the amount of information in a message and the loss of information when the message is transmitted. Here, the microstates typically are binary digits, or bits for the computer literate. In 1948, C. E. Shannon discovered that this formalism applied to communications. This was a revolutionary development that quickly led to a new discipline—*information theory*. This approach establishes a connection between information and entropy. Here is a very simplified illustration of the connection. Consider a highly disordered system, such as water vapor. The molecules are in constant motion and there is no information on their locations. Obviously, the entropy of the vapor is large. Now suppose the vapor is chilled so that the water molecules form ice. There is a great deal of information about the positions of the molecules in ice; they have to occupy certain positions in the ice crystal. Note also that the entropy of the ice is considerably less than that of the vapor. So the gain in information on the locations of the molecules is accompanied by a drop in entropy. Note that this loss of entropy does not violate the Second Law. Chilling the vapor implied a flux of energy and entropy out of the system to the environment. The measure of the information gained when the vapor condensed to ice is the difference between the entropies of the vapor and the ice. This connection has spawned the peculiar term "negentropy." In information theory, any gain of information about a system requires a loss of entropy from the system, or an increase in the negentropy of the system. A precise formulation of this idea is the "negentropy principal." This principle states that any gain in information of a system is less than the gain in entropy of the system *and its environment*. Although I think the term "negentropy" is confusing and unnecessary jargon, the negentropy principle is an important statement of the Second Law. It explicitly states that the flux of entropy between the system and its environment must be accounted for in any measurement. Applications and extensions of this concept have been made to such diverse disciplines as linguistics, telecommunications, signal processing, and computer codes.

The quote by C. P. Snow at the beginning of the Preface suggests that the notions of entropy, disorder, and the Second Law are deeply rooted in human culture. To illustrate this, recall the myth of the Greek goddess, Pandora, daughter of Hephaestus, one of the fire gods. According to the myth, Pandora opens the box and frees all of the woes that plague humankind. Opening the box is the same as opening the containers of CO_2 and N_2 in the gedanken experiment. The entropy of the world was increased.

Furthermore, this operation is irreversible; the woes cannot be put back in the box. The ancient Greek scholars were surely aware of MN2L!

The Classic Faces

The classic faces of entropy originated during the early part of the 19th century, largely due to the work of Sardi Carnot. His work dealt with the efficiency of an idealized engine, which we now call a Carnot engine. This engine uses gas in a cylinder to extract heat energy from a hot reservoir to perform work and then vents to a cold reservoir. The perfect Carnot engine would cycle indefinitely without any energy inputs other than from the hot reservoir. However, it was found that without additional energy input, the engine would not return to its initial state, even after one cycle. The reason is that energy is required to overcome friction as the piston moves in the cylinder, to open and close the gates to the reservoirs, and because thermal energy is conducted away from the cylinder to the environment.

There is a well-known expression for the macrostate efficiency of the Carnot engine. Letting T_1 and T_2 be the temperatures of the cold and hot reservoirs respectively, the efficiency is given by:

$$Efficiency = 1 - T_1 / T_2.$$

This formula shows that the efficiency of the Carnot engine increases with the temperature contrast between the hot and cold reservoirs. If the reservoirs are at the same temperature then the efficiency is zero. An efficiency of one means the engine converts all input energy from the hot reservoir to work. As the formula shows, this can only be achieved if the cold reservoir is at absolute zero.

What does the efficiency have to do with entropy? It is directly related to the ratio of the total entropy change to the entropy change of the cold reservoir. The entropy changes are always positive until both reservoirs come to the same temperature. Then both, along with the efficiency, go to zero.

There is also a Carnot refrigerator, and the expression for its efficiency is exactly the same mathematically. With a refrigerator, work *input* is used to move heat from the cold reservoir to the hot reservoir. Although it is impossible to build either a perfect Carnot engine or refrigerator, this formula is very useful as an upper bound for the efficiencies of real engines and refrigerators.

In addition to efficiency, another practical aspect of engines is the need to get rid of the combustion products, or exhausts. These products take the form of fumes or gases and solids. Just as our bodies must vent waste products when we work out at the health club, so must engines get rid of their wastes if they are to continue to operate.

The inability to construct 100% efficient Carnot engines led researchers to realize that this was due to a fundamental principle of nature. Now we call this principle the Second Law. One of the earliest statements of this principle was by Lord Kelvin, who said in effect that it is impossible to have a process whose sole function is to transform heat extracted from a source that is always at the same temperature into work. Rudolph Clausius, the same person who first coined the word "entropy," gave another version of the same principle. Clausius said in effect that it is impossible for any self-acting machine to convey heat continuously from one body to another at a higher temperature.

One product of the research during this period was a quantification of entropy. As with potential energy, it was the differences in entropy, rather than absolute values, that were important in these early applications. Thus, the first operational definition of entropy was given as the change in heat or internal energy a system experiences divided by its temperature, or $\Delta S = \Delta E/T$. The change in internal energy arises because of conversions of other forms of energy within the system and/or work performed by the system on its environment.

Just as there is a general statement of energy balance for a system, there is a statement for the entropy balance as well. In fact, the mathematical structure is exactly the same, even though the terms take on different meanings. The expression is:

$$\frac{\partial S}{\partial t} = T_S + G_S.$$

As in the energy balance equation given in the last chapter, the term on the left-hand side in the entropy balance equation is the rate of change with time of the entropy S of a system. The first term on the right hand-side, T_S, accounts for the transport across the system boundary. This includes both advection of entropy as well as entropy conveyed by heat and mass transported into or out of the system. As with the corresponding term in the energy balance equation in the last chapter, this term can be positive or negative depending on whether there is a net import or export of entropy through the system boundary. In an isolated system this term is identically zero. For closed and open systems T_S can be of either sign but is not exactly zero.

The source term, G_S, is the rate of production of entropy within the system. This can arise by diffusion of heat and different chemical species, mechanical dissipation from pressure and viscous processes, chemical reactions, and electric currents. Unlike the source term in the energy balance equation, this term is never negative. The conversion of energy to work by any system will *always* add to the entropy of the system whether it is open, closed, or isolated. This is the essence of the Second Law.

In elementary thermodynamics courses that focus on equilibrium conditions, T_S is identified as the heat flux across the system's boundary divided by the temperature of the system, and G_S is the pressure times the change of the volume of the system divided by the temperature. Recall from Chapter 2 that the flux of heat was included in T_E, temperature was one way of specifying internal energy, and the change in volume characterized the mechanical energy added to the system from the environment. The commonality of these terms, in both the energy and entropy balance equations, illustrates why MN2L should always be used together.

Practical applications are more complicated. Engines convert the chemical energy in fuel to work. The combustion products contain entropy, and so they contribute to the entropy production, G_S, within the system. Exhausting these products to the environment reduces the buildup of entropy in the system. Thus, T_S also includes entropy-containing exhausts. More generally, transformations of chemical energy to heat or other forms of energy always produce entropy-containing by-products. For example, smoking a cigarette converts the chemical energy in tobacco to heat. The waste products are smoke and residual, uncombusted compounds. In contrast to engines, it is the entropy-containing by-products of the combustion, and not the heat energy that is desired by smokers. These by-products have no energy or nutritional value for smokers, and so their bodies are not particularly efficient at breaking them down. Thus, they accumulate with potentially harmful effects.

In more general situations, where several subsystems are interacting, the processes contributing to the source term in the entropy balance can be grouped into the sum of the products of "fluxes" and "forces." The term "flux" is a bit unfortunate here since it does not refer to transports across the system boundary that appear in T_S. In the present context it refers to a primitive notion of cause and effect; the forces are the causes and the fluxes are the effects. For example, the flux of heat within the system (not into the system across the boundary) might be caused by the temperature gradient, that is, the change in temperature with position within the system. Another possibility could be the flux of chemical species within the system. This is caused by spatial differences (or gradients) in the concentration of the various chemical constituents.

The partition into cause and effect is a philosophical construct. If convenient, the roles of individual causes and effects can be reversed with no numerical effect on the entropy production. In essence, Mother Nature does not care about specific causes and effects when it comes to generating entropy. Her bottom line is that conversion of energy to work will always generate entropy products.

For specific applications scientists must establish empirical relations between causes and effects. Many of these have been so successful they have acquired the status of "laws." Examples readers may have heard of are: Fourier's Law, which states that the heat flux is proportional to the temperature gradient; Fick's Law, which says that the diffusion of a contaminant is proportional to the gradient of the concentration of the contaminant; the Navier Stokes Law, which says that the viscous stress in a flowing fluid is proportional to the velocity gradients in the fluid; or Ohm's Law, which relates the electric conduction current to the electric field. The fluxes in these laws are analogous to microstates in the probabilistic faces in that they involve processes that are not easily observed. On the other hand, the gradients are readily observed. Thus, these "laws" play a role similar to Boltzmann's equation, which relates the microstates to macrostates. With the traditional faces of entropy these empirical laws connect entropy production to microscale physics.

However, the ambiguity about cause and effect means that a heat flux, for example, could be caused by other mechanisms as well. Thus, the general theory had to allow for the possibility of "cross" effects. It turns out that these cross effects are not only observed, but also often have practical significance. One particularly noteworthy example is the thermoelectric effect. In the first part of the 19th century, it was observed that a voltage change across certain materials occurred whenever a temperature gradient was imposed. A dramatic example is when junctures of wires made of bismuth and antimony are immersed in a drop of water. Passage of current in one direction freezes the drop. Reversing the current then melts the drop. Lord Kelvin, the same person who gave one of the first statements of the Second Law, developed the theory for this in the mid-19th century. Applications of the thermoelectric effect are widely used in NASA space vehicles.

Characterization of the "cross" effects is of huge practical significance. At the same time, it deals with very basic theoretical and philosophical issues such as how microscopic reversible processes produce irreversible macroscopic processes. This involves theory that connects the probabilistic faces at a microscopic level to the classic faces on macroscopic scales. L. Onsager and G. Casimir, two Nobel Prize winners, are among many eminent scientists who have worked in characterizing "cross" effects in entropy production.

The reader should realize that during the 19th century there was an enormous amount of scientific and technical research on building bigger and more efficient engines. Moreover, the stakes were high. Not only were individual scientists competing against their peers for recognition, but there was also great competition in the private sector to access this technology for profit. National interests were at stake as well. Carnot himself recognized that France's technical and industrial inferiority vis-à-vis Great Britain contributed to her loss in the Napoleonic War. Along with fellow scientists and astute political leaders, he realized that nations possessing efficient steam power and access to coal would be military and industrial superpowers. These factors, perhaps more than anything else, powered science and technology during the 19th century. The Industrial Revolution and the recognition and quantification of MN2L are tightly intertwined. I believe that neither could have happened in isolation.

Aside on Carnot Engines

Virtually every undergraduate in physical sciences and engineering is expected to be able to describe the operation of a Carnot engine. Although the Second Law forbids the operation of such an engine, it illustrates nicely how a real piston engine works. Here is a brief account of the Carnot engine.

Think of the engine as a simple cylinder. Figure 3.1 illustrates schematically four stages of a typical engine cycle. Initially the gas in the cylinder is in thermal equilibrium with a hot reservoir. This is point A in the figure. Now the piston in the cylinder moves so as to expand the cylinder volume isothermally along the curve A to B. This expansion requires energy to move the piston and this comes from the extraction of heat from the reservoir. This part of the cycle is like the combustion of fuel in a real engine. At some point B in the expansion, the cylinder is disconnected from the reservoir so that no heat enters the cylinder. The expansion then continues adiabatically, meaning that no heat enters or leaves the cylinder. This expansion is indicated by the curve B to C. Continued expansion requires energy, and this comes from the internal or heat energy of the gas in the cylinder. From the simple problems in the last chapter, you know this lowers the temperature of the gas. At some point C the expansion stops and the cylinder is connected to a cold reservoir that is at the temperature of the gas in the cylinder. This part of the cycle is the start of the exhaust phase in a real engine. The piston then starts a compression stroke along the curve C to D. At first the compression is isothermal, which means heat flows from the cylinder to the cold reservoir at such a rate as to maintain a constant temperature in the cylinder. Then, at some point D in the compression cycle, the cylinder is disconnected from the cold reservoir and the compression stroke continues adiabatically until the temperature is raised to its initial value. At this point, the cylinder is connected to the hot reservoir and the cycle is repeated.

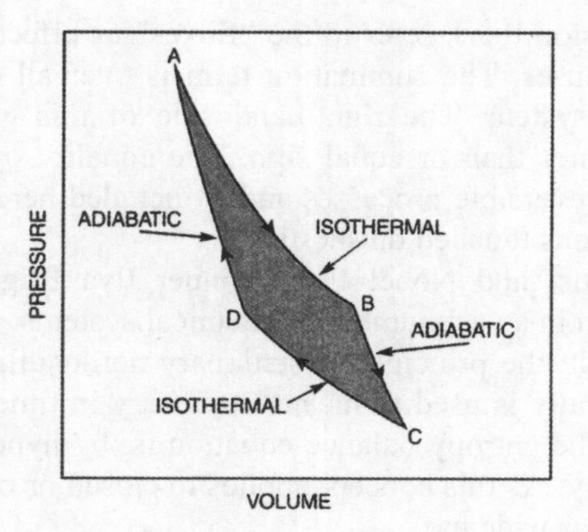

Fig. 3.1 Cartoon of a Carnot engine. See text for discussion.

Figure 3.1 captures the gist of real engines. The energy is delivered at high pressure and temperature (curve ABC) and the exhaust is at low pressure and temperature (curve CDA). The greater the pressure and temperature difference between the delivery and the exhaust, the greater the efficiency.

The efficiency of the Carnot engine also can be expressed in the language of the probabilistic faces. The macrostate of this system is the position of the piston. The microstates are composed of the gates to the reservoirs, miniscule dashpots accounting for friction, and photons radiating away heat to the environment.

Prigogine's Theorem

Here we want to quantify as simply as possible two critical aspects of the Second Law: the positive nature of entropy generation in a system and the decomposition of that effect into forces and fluxes, or causes and effects. The mathematical equivalency of this is:

$$G_S = \sum_i J_i X_i \geq 0.$$

In this expression the J_i refer to the "fluxes" or effects while the X_i are the "forces" or causes. The summation term is over all entropy producing processes in the system. The right-hand side of this equation shows an inequality, or greater than or equal sign. The equality option refers to the idealized case of reversible processes and is included here for historical and philosophical reasons touched on shortly.

Belgian scientist and Nobel Prize winner Ilya Prigogine has studied entropy production in a vast number of chemical systems. His research led to a remarkable result: the principle of stationary non-equilibrium. In science, "stationary" generally is used to mean "stationary in time." In this case the left-hand side of the entropy balance equation is, by hypothesis, zero if the system is stationary. As this concept applies to closed or open systems, there is great potential for wide use.

There are two consequences of stationary equilibrium. As $G_S + T_S = 0$ and G_S is positive, it follows that T_S has to be its negative, or $T_S = -G_S < 0$.

Thus there is a net flux of entropy out of the system in this case. The other consequence is that the production of entropy, G_S, is a minimum. It is stressed that G_S is a minimum but not zero in this case. Entropy is still produced!

There are a number of ways stationary non-equilibrium can be achieved (nearly) in practice. In Part 2, we will be particularly interested in the case where an open system receives some material M from the environment; the system transforms this material by chemical reactions to a number of intermediate compounds to produce a final product F, which is ejected back to the environment. Prigogine showed that if the amounts of the intermediate compounds are stationary, the entropy production by the system is a minimum.

An individual organism is an example of a system that is approximately in a state of stationary non-equilibrium. After it has reached maturity, the intake of food and nutrients is in mass balance with the excretion of metabolic wastes. When the organism dies, both the intakes of food and the excretion of metabolic wastes stops, but the organism continues to generate entropy through the decay of cells. Since the organism is no longer capable of exporting entropy, the condition for stationary non-equilibrium no longer holds.

A less gruesome example of the principle of stationary non-equilibrium is illustrated by an idealized closed ecosystem composed of two open subsystems. Figure 3.2 is a schematic of such a system. Being closed, the composite system must rely on photons or other ethereal energy sources and not directly on mass for energy. On the other hand, the two open subsystems are free to exchange mass, and hence energy, with each other. The plant subsystem plays the role of primary producer since it is capable of utilizing directly the energy in the solar photons entering the system to do work. By work I mean growth and procreation. The animal subsystem plays the role of primary consumer since it obtains chemical energy from the primary producer. As with the plants, this energy is used to perform work.

Both subsystems naturally obey Mother Nature's Second Law in that not all the energy consumed is available for growth and procreation. Some must be used for maintenance functions, foraging, defense, etc. Also, the conversion of energy to work itself is not perfect. Thus both subsystems excrete mass as part of their metabolic processes. This mass excretion, O_2 in the case of plants and CO_2 in the case of animals, carries away entropy harmful to the respective subsystem. Functionally this excretion is no different than the venting of fumes and ashes by a coal burning locomotive, and is equally important. The important point is that the O_2 is a metabolic fuel for the animals, while their CO_2 excretion serves the same metabolic function for the plants. Each group of organisms recycles the metabolic wastes of the other for the mutual benefit of both. If there were no animals, then the plants would soon choke on their own smog of O_2. Fortunately their smog is fuel for the animals.

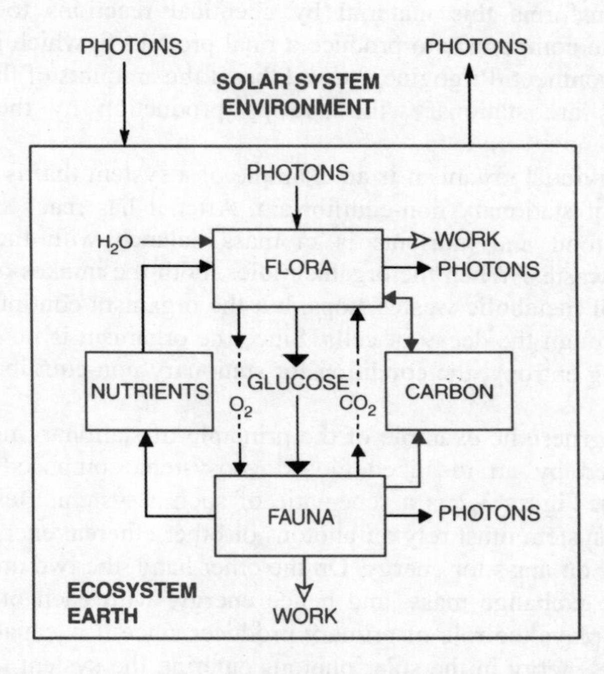

Fig. 3.2 Schematic of idealized ecosystem Earth. The blue and red lines depict energy and entropy flows, respectively, while the black lines show flows of water and carbon. Note that the O_2 metabolic waste of plants is fuel for animals while the reverse is true for CO_2.

As thermodynamic engines, neither plants nor animals are noted for efficiency; both produce a lot of entropy. The efficiency equation and the Carnot cycle demonstrate this. Both take in energy and expel entropy at nearly the same temperature.

Notice, however, that the animal subsystem uses the O_2 metabolic entropy product of the plant subsystem, and the CO_2 entropy product of the animal subsystem is used by the plant subsystem. If this mass flow between the subsystems is steady, then the idealized closed system is in a state of stationary non-equilibrium. This implies the system is producing the minimum amount of entropy even though neither subsystem is, and some entropy must be transported out of the system. Presumably this is carried away by expelled mass for open systems and perhaps photons for closed systems.

Of course, real ecosystems are much more complicated than this. There is an enormous hierarchy of smaller subsystems in both plant and animal subsystems along with convoluted energy and entropy connections. Moreover, this picture does not consider nutrients that are crucial for both plants and animals. It is interesting that there is a tendency towards a balance of nutrient flow between plants and animals similar to the metabolic balance of CO_2 and O_2. Generally, large reservoirs of nutrients as well as CO_2, O_2, and H_2O are needed to damp out fluctuations produced by different growth cycles of the subsystems. Microbial life forms and their attendant energy and entropy cycles are critical to the health of higher-level ecosystems, but these are not included here. Finally, all ecosystems are open unless one considers ecosystem Earth or the peculiar ecosystems found in caves, hot springs, or at deep-sea vents.

More generally, we think of primary producers as subsystems that convert primary sources of energy to work. In the example above, plants convert the energy of the sun received as photons to growth and regeneration. The primary consumer subsystems act as thermodynamic engines, which ingest and extract chemical energy from fuel; use this energy to do some work important to the subsystems like growth and procreation; and eject the combustion or metabolic products of the transformation processes. Now, if these products can be recycled through other types of subsystems, and if the overall concentration of these intermediate products is steady in time, then the entropy production for the system is a minimum!

There are several important lessons to be learned from this. It is important for subsystems to excrete their metabolic entropy products, and for these products to be used by other types of subsystems. If the flow of these intermediate products is steady in time, then the system produces the minimum amount of entropy. It is not important that the subsystems be minimum entropy producers. Finally, the system must export entropy products just as the subsystems do. All of this seems to be approximately true for stable or long-living ecosystems. Such systems are examples of nature's application of Prigogine's theorem.

In this discussion, I have deliberately used general terms like open and closed systems and subsystems in place of biological jargon. The beauty of the broad generality of this practice here, as in science, is that the same concepts are applicable to any number of situations. So even though the principle of stationary non-equilibrium is illustrated as an ecosystem, the same idea can be applied to other examples as well. In Part 2, we shall see how the principle of stationary non-equilibrium can be applied to some types of manufacturing.

Entropy and Change

Perhaps the most difficult aspect of nature we humans must deal with is change. Often devoutly desired, change is rarely appreciated when it occurs. Change is usually an indicator of the passage of time. Aging, for example, is a widely feared manifestation of change. More generally, we can retrace our steps to return to geographic positions, but once there we cannot go back in time to correct mistakes or relive prior successes. This is Thomas Wolfe's message in his book, *You Can't Go Home Again*.

The preoccupation with change rivals that with sex in the USA. The "anti-aging" industry generates billions of dollars annually for "cures" that have, at best, limited short-term cosmetic effects. Much of literature deals with aspects of humankind's love/hate attitude about change. Common themes are the inevitable forward march of time, the consequences of irreversible actions, and the occurrence of seemingly random events that alter the lives of the protagonists. Concern with change is so endemic to the human culture that it has drawn the attention of charlatans, who prey on the ignorant and fearful, as well as philosophers and scientists, who attempt to rationalize and quantify change for the rest of society. Part of the mystique of entropy and the Second Law is the subliminal belief that both are intimately linked to the changes we all experience during our lifetimes. In this context, entropy generation is a universal indicator of change. The three aspects of change most closely connected to the Second Law are technically referred to as the arrow of time, irreversibility, and causality.

Consider first the *arrow of time*. Basic laws of nature, such as Newton's law F=MA and Schrödinger's equation, which quantifies the rules of quantum mechanics, do not specify a direction for time when applied to elements of microsystems. In other words, perfectly valid (but different) solutions are obtained if time goes backward rather than forward! One solution moves forward in time (in sciencese this is called the retarded solution) while the other moves backward in time (or in sciencese, advanced solution).

To illustrate these solutions, consider the following gedanken experiment. Imagine sitting by a still pond. Now throw a pebble into the center of the pond. What you observe is a sharp depression in the surface as the pebble hits and sinks, quickly followed by an equally sharp geyser, and then followed by small waves propagating slowly outward from the point of impact. Every characteristic of these waves is described by the retarded solution to extraordinary accuracy. One important characteristic is that the wave train does not arrive at an off-center position until *after* the pebble hits; its arrival is retarded.

In contrast, the advanced solution propagates backward in time. If it existed, you would see the waves propagating inward towards the point of impact *before* the pebble landed, and converging to the geyser. Reports of this solution would not pass publication standards at *The National Enquirer*. So the situation is that retarded solutions perfectly describe radio waves, ocean waves, sound waves, and earthquake waves. The advanced solutions, which arise naturally from the very same models, don't even occur.

It is unsatisfactory for theories to predict phenomena that are not observed. On the other hand, the theories can't be completely wrong, since they predict observed phenomena to outstanding accuracy. I have been bothered by this ever since I was first exposed to electricity and magnetism as an undergraduate. It has bothered physicists and philosophers for a lot longer. This issue is also embedded in literature. Recall the white queen's admonishment to Alice, "It's a bad memory that only works backwards."

Much of undergraduate thermodynamics deals with reversible processes with special focus on isolated systems. Reversible processes are those in which the systems can oscillate between two or more states without any input of energy. The technical requirement for this is that $G_S = 0$. But as stated many times above, this can never be achieved exactly in practice. Another complicating factor, often obscured in many cases, is that systems cannot be isolated.

There are many examples of evolutions of systems that are nearly reversible. The motion of the Earth about the sun is one. Once a year, Earth returns to the same position relative to the sun. So in some sense, our motion about the sun is reversible, and it would be tempting to say that $G_S = 0$ for the Earth-Sun system. However, this narrow view neglects the irreversible effects of solar tides that, along with lunar tides, are slowly making days longer. All everyday phenomena are irreversible if one looks closely enough. Irreversibility means that as systems evolve in time they begin to exhibit some aspects of random behavior and forget previous states.

Causality refers to the primitive notion that there must be a cause for every effect. This notion is quantified in Newton's law, which treats force as the cause of the motion of bodies. The idea that any force acting on a body is the cause of the body's subsequent motion has been drilled into us since high school science. Interestingly, the forces may exist whether or not there are any bodies upon which they can act to produce events. Gravity is a prime example. It exists because of the masses of the Earth and some object of interest, and not just to make the object roll off a table and fall to the floor.

It should be pointed out, however, there are exceptions to the view that forces are causes. For example, if a particle with an electric charge is set in motion in a magnetic field, then its motion produces the "Lorentz" force that affects the trajectory of the particle. This force is not present if the body is at rest. Moreover, a moving, electrically-charged particle will induce its own magnetic field. Thus, a magnetic field can induce a Lorentz force to alter the trajectory of a charged particle, and the same particle can induce a magnetic field when it moves. Cause and effect can be quite murky.

Incidentally, the Lorentz force obeys a more general time reversal property than Newton's law of gravity. Consider the motion of a planet around the sun. The solution to Newton's law for this problem shows that if the sign of time is changed from plus to minus after some period of time, then the planet reverses course and returns to its starting position. If the same exercise is carried out with the Lorentz force, the charged particle will not return to its starting position. To reverse the particle's trajectory, it is also necessary to reverse the direction of the magnetic field, or equivalently change the sign of the charge on the particle.

Aside on the Arrow of Time, Irreversibility, and Causality

The Arrow of Time

Explanations for the arrow of time, are current "hot" topics of research among some physicists and philosophers. Several natural phenomena are commonly cited as characterizing the arrow of time. My favorites are:

RADIATION. Most radiation problems, such as waves on fluids and electromagnetic (radio) waves, are governed by equations that admit solutions that propagate in both directions in time. Predictions from these equations have been verified to exceptional accuracy. Yet one observes only outgoing waves after a source event. As noted above, if one drops a pebble in a pond the waves radiate away from the point of impact. The observed behavior can be explained precisely as a standard retarded solution to Newton's law as applied to hydrodynamics. A time-reversed or advanced solution, where the waves propagate backwards in time to the point of impact, also satisfies Newton's law exactly, yet it has never been observed. The same is true for electromagnetic or radio waves. Solutions in which the radar signal propagates back to the antenna are never observed. Only "retarded" electromagnetic waves are observed although the "advanced" solution that is reversed in time also satisfies Maxwell's equations.

Many noted scientists have offered technical explanations as to why the advanced solutions are never observed. For example, early in the 20th century, Walter Ritz and Einstein had a famous debate over Ritz's postulate of a "cosmological" condition that eliminated the advanced solutions. John Wheeler and his student R. Feynman (you met him in Suggested Reading from the previous chapters) also worked on this question and devised an "absorber" condition that eliminated the advanced solutions. Neither proposal, nor any other for that matter, has gained wide acceptance by researchers active in this area.

QUANTUM MECHANICS. Inferring a direction of time from quantum mechanics is also a formidable technical challenge since the governing principle, Schrödinger's equation, also admits time-reversed solutions. It turns out that the direction of time is established whenever a measurement is made. In standard quantum mechanics lingo, the wave function describing the quantum system "collapses" when a measurement is made.

THERMODYNAMICS. The classical thermodynamic explanation differs from the previous two in that solutions are not invariant with a change in the direction of time. This was recognized in the 19th century, and so the positive character of G_S became the first scientifically justified arrow of time. If one accepted the Second Law, then the evolution of isolated systems is constrained to be from low- to high-entropy states. Reversing time would require entropy to decrease, a phenomenon never observed.

Inferring a direction of time from the probabilistic face of entropy is a formidable mathematical exercise requiring considerable physical insight. The essence of the argument is that for isolated systems with large numbers of particles, it is exceedingly improbable for the system to evolve from one state to another with less entropy. It is much more likely that the system will evolve to a state with greater entropy. Perhaps the first to study this problem was Boltzmann, who argued that even though individual particles obey Newton's laws, which are reversible in time, the entropy of the system increases, and thus is not reversible.

COSMOLOGY. This arrow of time is connected with Einstein's general theory of relativity, black holes, and the "big bang." The direction of time may have been specified at the instant of the "big bang" when the universe was first formed. This arrow seems to be the most fundamental explanation of the one way direction of time.

When examined in detail none of these phenomenological arrows offers a completely satisfactory general explanation for the direction of time. Additional postulates are made, and there are subtle and tacit assumptions in the analyses. These combine to limit the generality of the conclusions. The point for the reader should not be the technical limitations on the generality of the arrows, but the fact that many independent lines of reasoning and observations indicate that time reversal and time travel are just science fiction.

These example arrows of time have been around for many years. Recently, Prigogine (his theorem on minimum entropy production was cited earlier in this chapter) proposed a new arrow of time based on notions of "complexity." His philosophical approach differs drastically from that of most scientists. As I understand it, his idea is, based on two underlying assumptions. First, instead of applying Newton's or quantum mechanics' laws of motion to individual particles in a system, as nearly all scientists do, and then deducing the system's statistical properties, Prigogine considers the statistics as fundamental. This finesses the issue of calculating the trajectories of huge numbers of interacting particles, but it raises the potentially more difficult issue of specifying the statistics a priori. The second assumption is that most, if not all, such systems are "unstable."

To avoid a highly technical discussion of unstable dynamical systems, readers might think of these systems as chaotic in the sense that the time evolution of these systems is hypersensitive to the minutest perturbations of starting conditions and parameter values. David Ruelle, in his book *Chance and Chaos*, gave one interesting example of such hypersensitivity. He calculated that the removal of one electron 10 billion light years away would alter the trajectory of an air molecule here on Earth, so that in just a fraction of a second, it would miss a collision with another air molecule.

Prigogine's ideas on the arrow of time are hardly in the scientific mainstream as of 2000. For one thing, with his approach, present conditions do not uniquely determine conditions at later times. Whether or not his ideas on the arrow of time become accepted depends on how they stand up to the scrutiny of the scientific protocol. This means Prigogine and his collaborators will have to publish solutions to significant problems in appropriate refereed journals. Moreover, his ideas will have to be applied by other researchers to other problems, and these results published in the refereed scientific literature. Finally, his status as a Nobel Laureate will have no bearing on the assessment of his research in this area. Clearly, it will take years before there is any consensus on his ideas. As Prigogine is over 80 he, like Boltzmann, may not be around for the resolution. The scientific protocol, and not opinion polls, litigation, ideological litmus tests, legislation, or media spin-doctors will determine the fate of this concept.

Irreversibility

The essence of irreversibility is that a system, once perturbed, cannot return to its previous state without exporting entropy or importing energy. However, under special conditions, the macrostates may occur repeatedly. It is appropriate to call such processes reversible, but with the tacit agreement that there has been some generation of entropy by the microstates that is not reflected in the macrostates. Of course, the macrostates may also undergo irreversible changes as well. A crude test of the latter is to make a video of the evolution of the phenomena and then run the video in reverse. If both versions seem equally plausible, then the process is reversible. A reversed video of the planets would show them rotating in the opposite direction while occupying the same positions seen in the original video. Another example is a rubber band. If one is stretched and released, it will oscillate indefinitely, and there would be no obvious difference between the forward and reversed videos.

In the case of the rubber band, reversibility is true only if we consider its macrostate. At the molecular or microstate level each stretch produces a change that is not completely reversed when the rubber band returns to its initial macrostate. Moreover, the changes in the microstates accumulate and ultimately will cause the rubber band to break if it is repeatedly stretched.

A more dramatic example is the sad story of Humpty Dumpty. Suppose a video was made of Humpty spinning on top of the wall. Conservatives might view the video and claim he is turning to the right while liberals could reverse the tape and claim he is turning to the left. In fact, no scientist can tell from the video which way he is truly spinning. The entropy of both solutions is the same, and either solution is physically plausible. Moreover, the right turning solution is simply the time-reversed version of the left turning solution. No matter which way he is spinning, after a while the spin will slow, he will wobble, and, according to the nursery rhyme, fall off the wall and splatter when he hits the ground. This is a catastrophically irreversible transformation from a state of low to high entropy. Not only can he not be restored to his original condition; but both liberals and conservatives would agree which version of the video had been reversed.

Causality

This is best described as the "principle" of cause and effect. The simplest version is the belief that every effect must have a prior cause. But when dealing with natural phenomena this version is far too naive. Not every single force or cause produces a unique effect. In reality, a cause may produce several effects. This is certainly the case with MN2L in that the transformation of energy produces work but also generates entropy. The example of the Lorentz force acting on a moving charged particle suggests there may be feedback between cause and effect. The moving particle could enhance an existing magnetic field. This would be an example of a positive feedback in which the effect reinforced the cause, thus enhancing the effect further. Of course, with a different orientation of the magnetic field the Lorentz force could produce a negative feedback in which the magnetic field generated by the motion of the particle would tend to cancel a preexisting magnetic field. Finally, if the charged particle is acted on by more than one force, such as gravity, then its trajectory (effect) is determined by more than one cause.

Another example of cause and effect is heat flux. Early in our lives we learned to associate the flux or transport of heat with temperature gradients. Thus, when it is cold there is a flow of heat from our warm homes to the outside. But is there a heat flux because of the temperature gradient between the homes and the outside or does the heat flux produce the temperature gradient? At the turn of the century theoreticians rather arbitrarily classified the gradient as the cause and the flux as the effect. The traditional version of the Second Law, however, doesn't require this distinction. As mentioned earlier, it is possible to transform these roles without affecting the value of the production of entropy.

These simple examples violate the notion held by some that cause and effect are uniquely related. That narrow interpretation of determinism is common with the political and media establishments, which often fail to recognize that simple technological "fixes" have the potential to produce undesirable effects.

When repeated experiments are possible, we become comfortable in assigning cause and effect even if we do not understand the process. We know when we turn the automobile ignition key the engine starts, although few can explain in detail all the processes that must occur for this to happen. We see an object because the retinas in our eyes sense photons emitted by the object. Computer screens are illuminated when electrons fired by an electron gun impinge on the screen causing photons to be emitted. An undesirable effect of the latter operation is that some electrons penetrate the screen, and may cause eye damage to the operator if basic precautions are not taken.

Establishing cause and effect is a great deal harder in disciplines where it is difficult or impossible to perform repeated experiments. In these cases, scientists must rely on statistical analyses of the data, as well as specialized theories, to infer cause and effect. The proper design of one-time experiments and data requires considerable planning, precise execution, and very careful interpretation. Even after all of this there is still some statistical uncertainty as to cause and effect.

Climatology is one area in which cause and effect are consistently misrepresented. During 1997-1998, media claims that El Niño was the "cause" of "anomalous" weather were particularly ludicrous. Equally disturbing were statements by politicians that this El Niño was the result of "global warming." Fortunately, comedians like Jay Leno and David Letterman introduced some sense of balance on this issue.

In fact, El Niño is just one aspect of a global-scale change in the atmospheric and oceanic circulation. There is a complementary condition called La Niña. The climate in the tropical Pacific oscillates between El Niño and La Niña. Both are associated with large-scale surface pressure, wind, and rainfall fluctuations in the tropical Pacific and Indian Oceans. These fluctuations occur over a time span of several years. The scientific name for this fluctuation is the Southern Oscillation, and the correspondence with El Niño and La Niña is referred to as the El Niño Southern Oscillation or ENSO. ENSO events have been occurring at least since the end of the ice age, and thus, are not directly related to recent concern of global warming.

ENSO is just one of a number of global climatological indices. It is worthy of special attention since there is some skill in predicting its onset with lead times of up to a year. However, to date, the ability to predict anomalous weather events elsewhere on Earth from knowledge of ENSO conditions has not been established. Outside the tropical Pacific, other climate indices are important, but as yet there is little or no skill in predicting their characteristics. The best that can be determined is a statistical correlation, that is, one part of North America may be unusually warmer or wetter, or there will be more or less hurricanes than normal. These correlations are not perfect since other climate indices also affect the weather in various parts of North America. They can act to enhance or minimize weather effects associated with ENSO. Finally, it should be kept in mind that these correlations are not predictions of specific weather events, nor can they alone establish a cause and effect relation between ENSO conditions and weather events.

As just noted, there is a correlation between weather patterns in many parts of North America and the occurrence of El Niño in the equatorial Pacific. But the storms that ravaged parts of the country during the 1997-1998 period were not generated in the tropical Pacific and did not move from there directly to North America, contrary to some news reports. Rather, these storms arose from the same atmospheric processes that have always produced mid-latitude weather. However, characteristics of the general circulation of the atmosphere do differ between El Niño and non-El Niño periods. El Niños have proven to be useful phenomenological harbingers of temporary changes in the general circulation of the atmosphere and ocean, which are the real causes of changes in weather patterns over North America. This raises the deeper question: What causes the changes in the general circulation during ENSO events?

ENSO and similar climate phenomena illustrate the pitfalls of attempting to assign simple cause and effect relationships to naturally occurring events. ENSO events involve complicated and poorly understood interactions between the atmosphere and ocean. These interactions are feedbacks that can act both to enhance and cancel ENSO events.

One of the most consistent misuses of science and technology by the media and political establishments is oversimplification of cause and effect. Be wary of arguments that rely on a few oversimplified cause and effect relationships with limited allowance for feedbacks.

Summary

Here is a summary of the main points discussed above. The Second Law is an important, but not universal, arrow of time; it is the explanation for irreversible behavior for macrosystems, but it makes no distinction as to cause and effect. The generation of entropy by macro objects is a well-established arrow of time as well as a cause of irreversibility. When friction is included in Newton's law, then time-reversed solutions to even simple problems exhibit a decrease in entropy with time. This friction also appears as a flux or effect in the entropy balance. However, it is not at all clear that friction is the basic reason Humpty was unable to return to his pristine state after he fell. Causality is a basic issue that science has not yet resolved. Individual causes generally produce more than one effect, and effects generally are connected to causes through feedback mechanisms. This tends to blur the distinction between cause and effect. Separating cause and effect requires input from dynamical theories, and perhaps new laws of nature. The Second Law, as understood today, is central to, but not the ultimate explanation of, the direction of time, irreversibility, and causality.

Violations of the Second Law

It is fashionable in some circles to claim exceptions to the Second Law. The essence of these arguments is that the generation of entropy by the system in question, G_S, is supposedly negative. I have heard many such arguments and "proofs," but know of no observations of violations of the Second Law. All the arguments/proofs I have been exposed to are flawed. These flaws can be broken down into the following categories:

a. The tacit assumption that the system in question is isolated when, in fact, it is closed or even open. This means that entropy is really being exported from the system, or mathematically, T_S is negative. Those who argue that biological growth and evolution violate the Second Law commonly make this mistake.

b. A tacit omission of critical processes in evaluating the source term, G_S.

c. A subtle change or inconsistency in the way entropy is calculated. This is often the case when the proponent claims to follow the probabilistic paradigm. One example of inconsistent terminology was noted by Grad in his analysis of the Gibbs paradox.

d. An ill-posed or nonphysical problem. Sometimes the proponent makes tacit or unphysical assumptions. In other cases, the argument is purely a mathematical probabilistic analysis dealing with the occurrences of rare events with no connection to natural processes.

Suggested Reading

A library search of "entropy" likely will produce a bewildering number of books and articles from many disciplines. Nearly all of these require at least a calculus background to be able to appreciate what the author is saying. I have read very few, but enough to warn the reader that some are a lot better than others. Nevertheless, it would be worth your while checking out some of the references that might surface from a random search. As my own tastes tend to favor technical accounts in quantitative disciplines rather than popular and metaphysical explanations, there is little guidance I can give the nonscientist. I hope those who wish to learn more about entropy will consider reading the introductions to some of the technical accounts. Here are some sources I recommend for further reading. This small list is hardly complete.

P. W. Atkins, *The 2nd Law, Energy, Chaos, and Form, Scientific American Library*, 1984. This is the best source of information on the Second Law for a nonscientist I have run across. Atkins is a physical chemist and his emphasis is on the probabilistic faces of entropy. However, he covers the traditional faces as well. The book is written at a level nonscientists can understand. The writing is clear, there are nice digressions into history, and many illustrative figures.

H. Grad, "The Many Faces of Entropy," *Communications on Pure and Applied Mathematics*, XIV, 323-354, 1961. Okay, this is a technical contribution; however, the introduction is worth reading even for nonscientists. Grad is an eminent scientist and mathematician who has made many fundamental contributions. He is noted for clear writing.

B. Hayes, "Debugging Myself," *American Scientist*, 83, 5, 404-408, 1995. The idea to use a computer to illustrate micro- and macrostates came from this insightful article. However, it is concerned with computer failures, not the Second Law.

H. C. von Baeyer, "Tiny Doubles," *The Sciences*, 37, 3, 11-13, 1997. This short article has a clever discussion about distinguishability that nonscientists should find enlightening. The author raises a provocative question: Why are everyday objects distinguishable when they are comprised of indistinguishable elementary particles?

N. Georgescu-Roegen, *The Entropy Law and the Economic Process*, Harvard University Press, 1971. This is an attempt to apply the probabilistic face of entropy to economics. Readers who claim some expertise in business and economics may gain some insight into this face from this book. If you are comfortable with logarithms and algebra, the math will not be a barrier.

I. Prigogine, *Introduction to the Thermodynamics of Irreversible Processes*, Interscience, 1967. This is one of my favorite books, and so I recommend it to everyone even though it is written for scientists who are not physical chemists. The focus is on the traditional face of entropy. The discussions are clear and simple, and instructive examples are given. One doesn't have to be a chemist to follow the text, but it helps to have a working knowledge of calculus.

T. Rothman, "Irreversible Differences," *The Sciences*, 37, 4, 26-31, 1997. This short but instructive, non-technical article gives some historical perspective to Prigogine's ideas on the arrow of time.

I. Prigogine, *The End of Certainty: Time, Chaos, and the New Laws of Nature*, Free Press, 1997. This is the most recent of Prigogine's books for nonscientists explaining his ideas about time's arrow. These books are quite popular with many nonscientists. However, in contrast to the technical reference cited above,

I find Prigogine's non-technical books written for nonscientists difficult to follow.

L. Brillouin, *Science and Information Theory*, Academic Press, 1962. This is a classic text on information theory written by one of its founders. Although it is technical, Brillouin explains carefully and simply the essential concepts. The interested nonscientist will learn a great deal from this book.

S. G. Philander, *Is the Temperature Rising?*, Princeton University Press, 1998. This book is written for nonscientists. Chapter 9 provides an introduction to El Niño and La Niña phenomena and the associated effects on weather.

M. Latif, D. Anderson, T. Barnett, M. Cane, R. Kleeman, A. Leetmaa, J. O'Brien, A. Rosati, and E. Schneider, "A Review of the Predictability and Prediction of ENSO," *J. Geophys. Res.*, 103, C7, 14,375-14,395, 1998.

P. J. Webster, V. O. Magana, T. N. Palmer, J. Shukla, R. A. Thomas, M. Yanai, and T. Yasunari, "Monsoons: Processes, Predictability and the Prospects for Prediction," *J. Geophys. Res.*, 103, C7, 14,451-15,510, 1998. Those interested in a recent assessment of prediction by scientists active in ENSO research would benefit greatly by reading these last two cited papers. Although they appear in a technical journal, readers should have no difficulty in understanding the gist of their findings just from the abstracts.

PART 2 CIRCUS EARTH

1. The Scientific Protocol and Critical Thinking

"Science is the organized systematic enterprise that gathers knowledge about the world and condenses the knowledge into testable laws and principles."

E. O. Wilson

Preview

Prior to the Industrial Revolution, science was the private domain of a few intellectuals often supported by wealthy patrons. The first scientific journal in Europe was the *Philosophical Transactions*, published in England by the Royal Society in 1665, about 200 years after the publication of the *Guttenberg Bible*. Within a few years, scientific journals were started in several other European countries. But by the start of the Industrial Revolution there were still only a handful of scientific journals, and the number of papers in each issue was small by today's standard. Many scientific results were communicated by letters and tracts. The Industrial Revolution changed this forever.

By the middle of the 19th century, the Industrial Revolution had matured to the point where it was transforming society. The sociological and economic consequences are well documented in many studies. But there is another aspect of this revolution that is not as widely recognized. This was a synergism between that revolution and a concurrent one going on in the sciences. The economic consequences of steam power motivated the governments, as well as large corporations in Great Britain, France, Germany, and even several smaller European countries, to invest heavily in scientific research, particularly in thermodynamics. Groups were started at universities that not only advanced knowledge through research, but also provided training for a new generation of scientists. One major culmination of this activity was the quantification of Mother Natures Two Laws (MN2L).

Although thermodynamic-based research and development received most of the financial support and drew the most public attention, there was a substantial spillover into all the sciences. In fact, many of the leaders in thermodynamic research during this period also made fundamental contributions in other scientific disciplines. The latter part of that century saw spectacular developments in such diverse disciplines as geology, biology, and medicine. Moreover, there was a dramatic rise in the number of scientific societies and scientific journals.

The popularly perceived economic benefits of basic research, government support, the increased number of formally trained scientists, and rapid growth of scientific societies and professional journals transformed the practice of science from a pastime to a profession. It is generally recognized that the vast sociological and economic changes during the 19th century were the consequence of the Industrial Revolution. It is less recognized that the evolution of science from a pastime to a profession was also a result of the Industrial Revolution. Because of the huge impact of MN2L on the Industrial Revolution, a case can be made that there was a synergistic relation between the Industrial Revolution and the rise of rational inquiry, critical thinking, and the evolution of the practice of science into a true profession. It seems appropriate then to pause a moment and review the elements of the practice of science or the scientific protocol.

There is another reason for this review. As stated in the Preface, rational inquiry and critical thinking are no longer in vogue in some intellectual circles. Furthermore, few outside science understand the scientific protocol. The primary purpose of this chapter is to expose readers to the scientific protocol, hypothesis testing, and some of the elements of critical thinking. The chapter ends with a fun application of critical thinking to the "Roswell Incident" and the popular issue of UFOs.

The Scientific Protocol

It is important to realize that MN2L did not result from some overnight brainstorm. In fact, it took nearly a century before the scientific community generally accepted the fundamental significance of these two laws.

Why did it take so long for these two fundamental principles to be recognized as such? The answer is that they first had to run the gauntlet of the scientific protocol. The first pillar of this protocol is the "scientific method," one aspect of a more general rhetoric that I call critical thinking. The philosophical roots of the scientific method extend back at least to the ancient Greeks; however, it was not until the middle 1600s that it emerged as a distinct rhetoric. No doubt, the widespread use of the printing press and the rise of literacy in the general population contributed to its development. Publication and widespread dissemination of scientific results were important media events of those times. The present version of the scientific method arose from efforts in the 1920s to codify all rational inquiry.

This method is a systematic procedure for evaluating competing explanations of observations. Here is how it might work in a hypothetical case: Suppose some scientists collect observations that do not fit into the prevailing paradigm. A number of hypotheses are formulated to explain the new results. The very first hypothesis is *always* experimental, procedural, or observational error. Often, apparent violations of MN2L are indicators of experimental errors. Once such errors are eliminated, then additional experimental and theoretical investigations are performed to assess the preliminary results. Presumably, the new round of experiments and theoretical studies eliminates some hypotheses, causes others to be modified in some fashion, and may even suggest new ones. Such probing analyses typically suggest additional experiments. This process continues until all but one explanation has been eliminated. Obviously, all of this takes considerable time. Years to decades are not unusual.

A few words about observations are appropriate. First, observations ideally should be made under tightly controlled conditions that can be repeated by other investigators. One-time only events cannot be studied using the scientific protocol. In fields like oceanography, atmospheric sciences, and astronomy, laboratory type experimental control simply is not possible. Fortunately for scientists (but not the population), natural phenomena such as tornadoes, typhoons, and earthquakes occur regularly. Although it is possible to observe these phenomena repeatedly, scientists in these disciplines cannot say exactly when the observations will be made, nor can they tweak the characteristics of the phenomena, as do their colleagues in the laboratory. Second, if two or more explanations tie in their ability to explain observations, the winner is the one with the least number of empirically determined parameters and restrictive assumptions. This aspect of the scientific method is an application of a general logical construct known as "Occam's razor."

At this point, the hypothesis may be elevated to a theory. But even a theory is open to further challenge. In fact, the fate of all theories is to be incorporated into or overthrown by new theories. The flux of half-baked ideas to hypotheses and ultimately to theories, which eventually are discarded, is what makes science exciting: You are never quite sure of the ground rules. It is also one of the aspects of science most misunderstood by nonscientists, who seem troubled by the continual questioning of scientific results. They may prefer to think scientific theories should be chiseled in stone like the Ten Commandments. Others may interpret the continual questioning as indicative that all science is wrong and should be ignored. Of course, both beliefs are wrong.

This notion of continual testing of theory to refute, modify, or generalize is the essence of the scientific method. Implementation of the method is based on two covenants. The first is open exchange of data. After the data gatherers examine the data and prepare analyses for publication, the data are made available to other interested scientists for independent analyses. The other covenant is publication of results in established, peer-reviewed scientific journals. Publications must document experimental or analysis procedures in detail sufficient for other scientists to reproduce the experiments or analyses.

Here is how peer review might work in the hypothetical case. After they are sure of their results, the scientists draft an article describing the findings of a phase of their research. Often, but not necessarily, the draft is circulated to other scientists for comment. This is done with the understanding that the material in the draft is "privileged" and should not be divulged without express approval of the authors. The original draft may be substantially modified in response to the informal criticisms of colleagues. I know of cases where the informally circulated draft was so substantially modified that new authors were added, as well as cases where the original version was set aside and never submitted. It is simply wrong to think that drafts or technical reports are true sources of scientific results or that the reviewed articles are, in some way, corrupted. The conclusions expressed in drafts and reports generally are not the authors' final conclusions.

At some point the draft is submitted to an appropriate scientific journal for peer review. A standard policy of these journals is that the results be significant and not previously published. The editor of the journal selects reviewers who are knowledgeable on the topic of submission. The reviewers act as unpaid consultants to the editor. Sometimes the reviewers may have also commented on the draft version, but this is not always the case. The number of reviewers varies with each journal, the nature of the paper, and the editor's whim. Two to three reviewers are typical, although I once submitted a paper that had seven different reviewers.

The reviews are anonymous and every aspect of the paper is fair game. The reviewers comment on the validity of the assumptions made in the study; the experimental and/or analysis procedures; the review of prior work on the topic; the significance of the findings; the quality, quantity, and appropriateness of the figures; and even the syntax, grammar, and spelling. Sometimes there are personal barbs in the reviews. Authors and editors must ignore these and concentrate on all technical deficiencies noted in the reviews regardless of their tenor. Based on the reviews, the editor decides to accept the paper as is (extremely rare), accept with modest modifications (usually grammar and spelling), suggests the authors prepare a revision that addresses the reviewers' concerns, or rejects the paper. The revision may go back to some or all of the original reviewers or to all new reviewers. Based on the second round of reviews, the editor usually makes the decision to publish or reject the submission; however, multiple rounds of review are not uncommon.

The editor is not obliged to go with all or even any of the reviewers' recommendations. The editor should review the reviews and decide if the criticisms are sufficiently justified. During my tenure as editor of the *Journal of Geophysical Research*, I rejected a paper that the reviewers liked, and accepted two that had negative reviews. It should be obvious from this that peer review is not a democratic process. The editor is truly a tsar and ultimately his/her decision stands. As was the case with real tsars, if the paper turns out to be wrong it is not the editor or reviewers who are embarrassed.

But the peer review story does not end here. Once the paper is in print it is still fair game for criticism. Other scientists may submit commentaries on the publication. These range from supporting data or analyses to devastating refutation of the findings. It is important that the commentaries be reviewed. This keeps poorly thought out or inappropriate criticisms from having a deleterious effect on a meritorious contribution, and it also prevents personal feuds from making their way into the scientific record. The authors are given a chance to reply to the commentaries. Of course, the replies are also reviewed.

A typical, but by no means universal, period for which commentaries are considered for publication is one year from the date of publication of the target paper. However, authors are not off the hook after one year. Subsequent contributions may embarrass the authors by presenting new findings or new analyses that invalidate their study. In fact, some of the most heated controversies in science have played out in a series of papers extending over several years.

Peer review does not get the public attention and respect it truly deserves. It is generally ignored in popular books on science, even by famous scientists, and only cursorily mentioned in books or articles on famous scientists. With the exception of a few health-related papers in journals such as the *New England Journal of Medicine* or *Science,* most news accounts and government releases do not cite reviewed articles. Peer review also is under attack by "revisionists" who claim it only serves the interests of establishment scientists and that it all too often is used to delay or even thwart publication of competitive approaches. Pay no heed to the naysayers. Peer review is alive and well and not in danger of extinction.

Peer-reviewed publications are not infallible. They are, however, considerably more reliable than results that have not gone through this ordeal. For consumers of scientific research, which we all are at some level, peer review means the results are suitable for additional evaluation in the scientific marketplace. This is not unlike beta testing of new products. Keep in mind that peer-reviewed papers are fair game and anyone is free to use or challenge the results as they see fit. CAVEAT EMPTOR always applies to peer-reviewed articles.

Media reports on science focus disproportionately on books and presentations at scientific meetings. In some scholarly disciplines, books and lengthy monographs are the preferred method of measuring achievement. This is not the case in science. Although very important in the codification and dissemination of scientific knowledge, books simply do not have the level of peer review that journal articles do. Books generally are used as texts or to summarize a mature line of research, not to present new findings. The level of review of papers and talks given at meetings is nowhere near that of journal articles. Such presentations usually discuss work in progress, subject to later revision, before submission to journals. Nevertheless, abstracts of papers presented at meetings often are not peer-reviewed but may be subject to criticism in the reviewed literature. This and the compression of time are the chief reasons most meeting abstracts are so vague and bland.

The point of this is that the only statements of record for scientific results are peer-reviewed articles. Hence, nonscientists should generally ignore any scientific claims that have not appeared in peer-reviewed journals, even when they are published by otherwise reputable elements of the print media or reported on national network evening news segments. Also, preprints and unpublished manuscripts are privileged information and should not be widely disseminated. In fact, publication of findings in a newspaper or any public forum may disqualify scientific articles from appearing in refereed journals since all journals require the material being published to be new and not previously published. So, nonscientists should not waste time pondering the significance of reports that have not been peer-reviewed no matter how much hype they get from the media or political establishments. The scientist's version of "where's the beef?" is "where's it published?" Remember, even after the results appear in peer-reviewed journals, "ignore" can be upgraded to "suspect" until further confirmation from other scientists working in the same area is also published.

The message here is simple and direct: *Readers are well advised to discount all news reports of scientific breakthroughs based on internal or technical reports, draft manuscripts, books, and meeting presentations.*

The scientific method, free exchange of data, and peer review constitute the scientific protocol. It is presumptuous to think that we mortals are capable of learning all of Mother Nature's secrets; hence, science is incomplete. For this reason, scientific theory is always subject to modification, and even repudiation based on further observations and analysis. In this way, knowledge, understanding, and technical progress are achieved. This protocol is what separates science from religious views requiring absolute dogma, politics where scientific and technical policy matters may be biased by PAC support, and media news often based on rumor or sensationalism.

The practice of science is the application of the scientific protocol. When I was a graduate student at Texas A&M University, an eminent scientist, Carl Rossby, visited the campus. In one of his lectures, he gave what I consider the most succinct description of the practice of science I have ever heard: "Science is the replacement of big errors by lesser errors." Readers would do well to keep this in mind when evaluating reports of scientific developments.

Aside on Hypothesis Testing

A critical factor in the acceptance or rejection of a theory is an assessment of how well it explains experiments or observations. This is one of the most difficult and fascinating aspects of science. The assessment must take into account experimental or observational errors, as well as the efficacy of different competing explanations. To do this objectively, scientists often rely on a methodology developed by statisticians. A generic name for this is hypothesis testing. Hypothesis testing is an important aspect of the scientific protocol. As one of the most widely used statistical procedures its effects are pervasive throughout society. Yet, it is something most nonscientists know little about. One facet of science consistently misreported by the media and often misinterpreted by politicians and even business executives is the implication of uncertainties in experimental results. The situation is particularly egregious in health-related matters, but distortions exist in all scientific and technological disciplines. Readers may be interested in learning a bit about what is really going on.

Broadly speaking, uncertainties in experiments can be divided into two types. One is random error in the observations; the other is ignorance about the true condition of Mother Nature. Obviously, the former affects inferences about the latter; however, the principal concern of readers is the latter category. The simplest situation to consider is whether a conjecture or hypothesis concerning Mother Nature is or is not true. It is helpful at this point to revert to sciencese and call H_0 the hypothesis that Mother Nature is in some specific state. Often, this is referred to as the null hypothesis. In order to make an objective decision as to whether H_0 is or is not true; the investigator needs appropriate criteria. Keep in mind, however, the actual state of Mother Nature isn't known until after the fact. Thus, whenever a hypothesis is accepted or rejected, there is the possibility of an error. If the investigator accepts or rejects H_0, and this turns out to be the correct decision, then the investigator deserves a pat on the back for being so clever. Unfortunately, the investigator is not always right. In this case, two types of errors in judgement on the true state of Mother Nature are possible. Suppose H_0 is really true but the criteria selected cause it to be rejected. Then a type I error, rejecting a true hypothesis, is made. The other judgement error occurs if H_0 is really false but the scientist elects to accept it as true. This is called a type II error, accepting a false hypothesis. Criteria for acceptance or rejection of hypotheses should be stated at the beginning of the experiment but often are arbitrary. If sufficient observations are available, significance levels of 0.05 or 0.01 are commonly used. These numbers give the percentage that the investigator will make the appropriate type of error.

In practice, great care is taken in formulating the exact wording of the null hypothesis, as well as the decision whether to test if it is true or false. Generally, the significance levels are not the same, nor are the consequences of making the two types of errors. Accepting a false hypothesis usually has different statistical significance and scientific consequence than rejecting a true hypothesis. These, and many other factors, enter into decisions on precise wording of the hypothesis, as well as which type of error is the least damaging. Depending upon the nature of the experiment and the consequence of the two types of errors, more importance may be attached to the rejection of H_0 than its acceptance!

In any set of experiments there are statistics associated with the observations that are "controlled" or otherwise specified from theory or other considerations. These statistics tend to be small when H_0 is true. In the case of just two possible states of nature, we can restrict attention to just one statistic. Thus, if this statistic exceeds some critical value, then H_0 is rejected. Varying the controlled critical value generates a critical region. Once all this is set up, standard statistical machinery determines the significance level of a type I error. When the significance of the type I error is determined, it is possible to calculate the significance level for the type II error. Predictably, if the controllable statistic produces a small chance of making a type I error, then it will also produce a larger chance of making a type II error. Essentially, the scientist cannot have it both ways and the reader should be aware of this basic asymmetry in statistical hypothesis testing.

Here is how all this might work in the purely hypothetical situation, of testing a drug to see if it can cure the common cold. A moment's reflection indicates the complications in assessing such a drug. If a thousand adults are exposed to a cold virus, then some percentage will actually catch a cold. All who catch a cold should recover in a few days whether or not the drug is effective. Moreover, some of those who catch colds may exhibit improvement even if they are given sugar pills instead of the real thing. This is known as the placebo effect, which incidentally works both ways. Some test subjects may develop cold symptoms even if they are not really exposed to the virus. Finally, in a large population, some percentage may actually catch a cold even if that population is not exposed to the virus, and the individuals do not exhibit the placebo effect.

Now suppose scientists at a pharmaceutical house are working feverishly on a drug to cure the common cold. Being first not only will bring instant fame and a Nobel Prize; it will make a huge amount of money for the scientists and the stockholders. After years of painstaking research the scientists have developed a drug that looks promising, so they start testing its effectiveness. Do they test the hypothesis that the drug is effective in curing the common cold? Not likely, since the legal and moral ramifications of accepting a false hypothesis that the drug is effective when it really is not (type II error) could be worse than rejecting a true hypothesis (type I error). A better strategy would be to test the hypothesis that the drug is not effective! If this hypothesis is rejected at some significance level, then the pharmaceutical house publicists can announce confidently, and with a small possibility of making a false claim, that the drug is effective. On the other hand, if the hypothesis that the drug is not effective is accepted then no announcement of a medical breakthrough is made, and the scientists continue their quest for fame and fortune back in the laboratory. So whenever the media announce a great medical breakthrough, remember this likely is based on minimizing a type I error and that there is a consequent greater probability that a type II error was made. Presumably, the formulation of the hypothesis is such that the consequences of the latter are not severe.

Screening tests used for certain types of cancer, cholesterol, substance abuse, etc., have the potential for confusion. A common misconception is that these tests are infallible yes/no types of tests. Although these tests generally are quite good, they are not infallible. Nor do they give yes/no types of answers. In fact modern technology is capable of giving a complete spectrum of readings for nearly all of these tests, although for cost reasons this is not always used. In the case of cholesterol, for example, the report gives numbers for the amounts of different types of cholesterol. But in other types of tests the subject is advised only that they tested positive or negative. Presumably the hypothesis being tested is that if a priori prescribed test readings are reached, then the subject is in the "red zone" and some remedial action is required. This suggests the test procedure is set up to minimize rejecting a true hypothesis, or a type I error.

However, a host of factors such as calibration drift, interactions of irrelevant chemicals in the body confusing the test, prescription drugs, and even operator error can occasionally conspire to produce results that exceed threshold values. In other words, minimizing a type I error ensures in these cases a greater chance for a type II error. If this is the case, the subject is the victim of a false positive. The consequence of a false positive for the victim can be quite serious. It could result in further expensive testing and, in the case of controlled substances, immediate dismissal. Moreover, the agency sponsoring the tests does not get what they paid for. Depending on the circumstances these agencies may incur additional expenses associated with the loss of time, loss of productive employees, or even litigation.

A better course of action in such cases is to use a method known as sequential analysis. This is a modification of the two-hypothesis test just described. With this method, the critical region for the test statistic is divided into three segments instead of two, and two prescribed critical values are used instead of one. If the test results are less than the lower of the critical values, then the subject passes. But if the results exceed the larger critical value, then the hypothesis is accepted, and the subject fails the test. Finally, if the test results fall between the two critical values, then another test is performed or additional data are collected. This method is commonly used in product testing because it has proven to be more cost effective than the two-hypothesis test with an a priori specified sample size.

In epidemiological studies even more sophisticated statistical methods are required. Here, a sample population of individuals is divided into two sub-populations. The sub-populations should be identical in every respect except that one was exposed to a disease or underwent a certain medical procedure, while the other sub-population did not. The sub-populations are studied to determine if any statistical difference in health related issues develop over time. A short reflection on potential problems with this approach should convince readers this is not an ideal way to establish a causal link. But alternatives, such as deliberately exposing a portion of the population to a disease or an untested and potentially dangerous procedure, are unacceptable.

Astute readers realize that Mother Nature rarely can be restricted to just two conditions. In fact, there usually are several likely states to consider. Not to worry, statisticians have been able to extend the approach described above to test any number of conditions. Such tests go under the generic name of *multiple decision methods*. Clearly, the logic that goes into formulating the hypotheses for these methods is much more complicated than the simple examples discussed here.

To get a driver's license, one merely needs to demonstrate elementary driving skills and a rudimentary knowledge of traffic laws—not a degree in mechanical engineering. Analogously one does not need a degree in science or statistics to understand the implications and uncertainties associated with hypothesis testing. In modern society, both a driver's license and a nodding acquaintance with hypothesis testing are useful.

Critical Thinking

A number of commentators have noted a general degradation of our ability to think critically. Critical thinking, like the martial arts, is a hard skill to acquire. Years of practice in different contexts and disciplines are necessary to reach a reasonable level of competency. Moreover, skill in critical thinking is not always encouraged or appreciated in the workplace or in personal relationships.

Superior critical thinking has been a traditional American characteristic since the founding of the Republic, and no doubt this trait played an important role in the ascendancy of American economic power. Today, politicians often regard scientific and technical issues as pawns in ideological debates while the media often sensationalize dubious scientific claims and trivialize fundamental developments.

There is an equally important, but more subtle, role for critical thinking by the public. Science and technology now provide an impressive array of conveniences and personal options. The technology is available to send people to any planet in the solar system and to extend life indefinitely through life support systems and organ transplants. Moreover, the media is rife with promises of even more dramatic developments regarding communications, computers, and genetic engineering. Some of these technologies should be pursued vigorously; others will cause serious long-term societal problems. Rather than passive reliance on political action committees, critical thinking should be used in making decisions about which technologies to pursue.

The only exposure to critical thinking many readers have had likely came through science courses where some mention of the scientific method was made. In essence, the practice of science is an exercise in critical thinking. A. B. Arons correctly noted that ideas come first and technical details afterwards. Unfortunately, the tendency today is to emphasize the latter at the expense of the former in many of the required science courses for non-science majors.

I try to use critical thinking in science as well as in my everyday struggle for existence. Arons has provided a list of 10 ingredients to critical thinking (*Teaching Introductory Physics*, A. B. Arons, "Chapter 13: Critical Thinking," Copyright 1997, by John Wiley & Sons, Inc. Reprinted by permission of John Wiley & Sons.). It is the best I have run across. Below is an abridged version of his original list. This is a poor substitute for the original. My wording is given in brackets.

1. Consciously raise questions like "What do we know? . . . What is the evidence for? . . . Why do we accept or believe? . . ." [when confronted with a problem or decision.]

2. Be clear and explicitly aware of gaps in available information. Recognize when a conclusion or decision is reached, [it usually is done so in the absence of complete information. Be prepared to tolerate the attendant ambiguity, uncertainty, and consequences.]

3. Discriminate between observation and inference . . . [fact and conjecture.]

4. Recognize that words are symbols for ideas and not the ideas themselves. [Be alert to situations where the same words are used as symbols for different ideas.]

5. Probe for assumptions [particularly implicit and unarticulated assumptions] behind [all] lines of reasoning. [In common vernacular, always be on the lookout for a hidden agenda.]

6. Draw inferences from data and other information you have faith in. Recognize when firm inferences cannot be drawn.

7. Given a particular situation apply relevant knowledge, experience, and plausible constraints to visualize . . . likely outcomes that might result from . . . [possible changes or incomplete knowledge of the situation.]

8. Discriminate between inductive and deductive reasoning; that is, whether the argument [extrapolates from the particular to the general or interpolates from the general to the particular. Extrapolation is riskier than interpolation but often is unavoidable.]

9. Test [all] lines of reasoning and conclusions for internal consistency. [I find it helpful to try to apply all lines of reasoning to special cases where I believe I know what the answer should be. Be wary of lines of reasoning that do not give consistent answers.]

10. Stand back and recognize the logical processes being used. [Are they appropriate for the situation at hand?]

Here are two other principles I have found very useful in critical thinking:

11. Correlations alone do not establish cause and effect. This is particularly useful in assessing media and government reports on health issues.

12. Assess the consequences of accepting a false hypothesis and rejecting a true hypothesis from available data. Be aware that the attendant consequences as well as the statistical uncertainties in the two cases often are vastly different.

An Application of Aron's Tenets of Critical Thinking

With apologies to the Roswell Chamber of Commerce and just for fun, let us examine the Roswell Incident in light of the rules of critical thinking. The Roswell Incident was one of the great cultural and media happenings of the latter half of the 20th century. Its origin goes back to July 1947 when a foreman on a ranch near Roswell, NM came across some strange debris. He turned the material over to the local sheriff who, in turn, turned it over to the Air Force (then a component of the Army). After examination, the military announced the material was nothing more than a downed weather balloon. The incident was more or less forgotten by the media for about 35 years although right from the gitgo there was a small, hard core group that claimed a conspiracy by authorities to cover up the "truth." It was revived when several pseudo-scientific books on the incident were published in the early 1980s. These books contended that the debris was really a crashed UFO, and that alien bodies had been recovered and are interred in the notorious "Area 51." Since then, there has been a media frenzy. The incident is the basis for a popular TV show, "The X-Files," and is a recurring theme in several popular movies. In July 1997, over 100,000 people came to Roswell for the 50th anniversary of the alleged UFO crash. As a source of entertainment, the Roswell Incident is superb. But as a real UFO crash, the information cited by adherents does not stand up to critical thinking. Here is my assessment of this situation based on Arons' tenets of critical thinking as applied to the hypothesis "did aliens crash at Roswell?"

1. Consciously raise questions like "What do we know? . . . What is the evidence for? . . . Why do we accept or believe? . . . when confronted with a problem or decision."

As far as I can tell, all that is really known about this incident is that the Air Force regularly released weather balloons in this area during this period and that photographs of what are claimed to be alien bodies have appeared in a number of places. To my knowledge, these photographs have never been authenticated. There have been many reports of UFO sightings worldwide. Some even claim evidence in historical records of sightings as far back as the Middle Ages. But apparently there is no other site with the mystique of Roswell.

2. Be clear and explicitly aware of gaps in available information. Recognize that when a conclusion or decision is reached, it usually is done so in the absence of complete information. Be prepared to tolerate the attendant ambiguity, uncertainty, and consequences.

The gaps in available information are huge. Details on the construction and propulsion of UFOs are not available. Nothing is known about how UFOs communicate with each other or even where they might come from.

3. Discriminate between observation and inference, fact and conjecture.

The lack of hard physical evidence for public examination restricts inferences to: 1) the aliens possess "Star Trek-like" technology; 2) there is a massive conspiracy to cover up the crash and what is known about alleged alien technology; and 3) there was no crash.

Note that the first two inferences are not mutually exclusive.

4. Recognize that words are symbols for ideas and not the ideas themselves.

There seem to be two ideas fueling the interest in UFOs. One is a massive conspiracy to cover up what is actually a nice thought: There are sentient beings outside planet Earth. Moreover, these beings are technologically superior to us. UFO sightings and the Roswell Incident provide the gist for the verbiage that keeps this thought alive. It is important to point out here that even if UFOs are not carrying ETs and there was no crash at Roswell, the idea of sentient beings elsewhere in the universe is not invalidated.

5. Probe for assumptions [particularly implicit and unarticulated assumptions] behind all lines of reasoning. [In common vernacular always be on the lookout for a hidden agenda.]

The people who push the Roswell Incident and UFOs on their followers have a failsafe strategy—conspiracy. Conspiracy theory can mold any fact, either for or against, into conformance with the conspiracy. Moreover, appeal to "Star Trek" technology can explain anything. "Star Trek" technology obviates any questions concerning propulsion and communications. Thus, the leaders in this movement, many of whom make a living at this, never have to produce a proverbial smoking gun.

There is another facet to this that might be pondered. Many UFO advocates tacitly assume the aliens are benign and intend to do us no harm (other than to poke, prod, and perform sexual experiments). History suggests this is not likely to be the case. Every instance when a technologically superior culture encountered a technologically inferior culture resulted in enslavement, exploitation, and devastation for the latter. On a more basic level, there are many examples of new species invading stable ecological communities. The inevitable result was annihilation of competitive existing species, and ultimately, a new ecological equilibrium. If this is a universal biological principle, then we should all hope aliens with technology vastly superior to ours do not visit Earth.

6. Draw inferences from data and other information you have faith in. Recognize when firm inferences cannot be drawn.

For those who accept "Star Trek" technology and conspiracy cover-up on faith, there is no doubt their inferences are not based on tangible data. For the rest of us, it is clear that, until physical evidence is made available to scientists for examination, there simply is not enough information to make a decision about the existence of ETs and UFOs.

7. Given a particular situation apply relevant knowledge, experience, and plausible constraints to visualize . . . likely outcomes that might result from . . . [possible changes or incomplete knowledge of the situation.]

From what we know about our solar system and its neighbors, any UFO would have had to travel many light years to get here. Arriving here in a reasonable time would require traveling greater than the speed of light or traveling through wormholes. Relative to our present technical capabilities, this really is science fiction.

8. Discriminate between inductive and deductive reasoning; that is, whether the argument extrapolates from the particular to the general or interpolates from the general to the particular. [Extrapolation is riskier than interpolation but often cannot be avoided.]

Those who believe in UFOs attempt to extrapolate from a dubious crash to a generality concerning life elsewhere in the universe.

9. Test all lines of reasoning and conclusions for internal consistency. [I find it helpful to try to apply all lines of reasoning to special cases where I believe I know what the answer should be. Be wary of lines of reasoning that do not give consistent answers.]

Conspiracy theories and "Star Trek-like" technologies are generally internally consistent since any inconsistency is part of the conspiracy, or just an aspect of unimaginable technology. However, the logic of UFO enthusiasts is not internally consistent. To illustrate this, assume that ETs exist and possess "Star Trek" technology. Why then would they crash? And if they did crash, why didn't they mount a rescue effort? Many UFO advocates do not claim the ETs are interfering in our lives. This suggests they adhere to something akin to the Prime Directive supposedly followed by The Federation in "Star Trek." If this were the case, then they would mount an extensive rescue effort. Finally, it could be argued that ET technology is good, but not good enough to eliminate accidents. If that is the case, then why would they send manned probes on such a sensitive and dangerous mission when they could recover all the information about us from unmanned probes?

10. Stand back and recognize the logical processes being used. Are they appropriate for the situation at hand?

Perhaps UFO advocates should consider the hypothesis that a crash really did not occur at Roswell.

Finally, consider the wide coverage recently given to the Roswell Incident in a broader context. A very small percentage of people claim to have had contact with extraterrestrials. Many more, but still a very small percentage, have observed unexplained phenomena in the atmosphere and are convinced that their experiences are indicative of extraterrestrials. It is not appropriate here to question the sincerity or sanity of these individuals. By most accounts these people believe extraterrestrials are benign or even friendly. Some even feel extraterrestrials will intervene to save us from ourselves. These are all characteristics of religious beliefs. Their observations and experiences thus could be termed miracles. So perhaps the Roswell Incident should be regarded as a religious movement. Those who chose to believe in UFOs and extraterrestrials are merely exercising their First Amendment rights. In this context, the recent gathering at Roswell is no more worthy of news coverage than the convention of any traditional religious group or denomination where a comparable number of participants attend. One wonders why the former draws much more media attention than the latter.

For the critical thinker the issue here is not whether one believes or does not believe that aliens crashed at Roswell. The issue is "where's the proof?"

Suggested Reading

P. W. Atkins, *The 2nd Law*. Scientific American Books, 1994. Chapter 1 has a nice synopsis of scientific developments during the 19th century that led to quantitative statements of MN2L. This includes biographical sketches of the scientists generally credited for these developments as well as commentary on the role of these laws in the Industrial Revolution.

A. B. Arons, *Teaching Introductory Physics*. Wiley and Sons, 1997. Do not let the title of this book throw you off. If I were tsar, I would require all members of Congress and newspersons to pass written and oral tests on Chapters 12 and 13 in Part I: "Achieving Wider Scientific Literacy" and "Critical Thinking," respectively.

The Flight from Science and Reason, edited by P. R. Gross, N. Levitt, and M. W. Lewis, Annals of the New York Academy of Sciences, vol. 775, 1996. This book provides many examples of the assault on critical thinking and science by "revisionists," many of whom are academics. Most of the articles are accessible to nonscientists. Some of the examples given here are frightening.

R. E. Strauch, The Fallacy of Causal Analysis, Report P-4618 from the Rand Corporation, 1971. If you are interested in learning more about the difficulties in drawing inferences from purely statistical analysis, then read this short tract. Strauch gives one of my favorite examples of erroneous conclusions based on correlation. The correlation of the word "cardinal" between the St. Louis baseball team and officials of the Catholic Church does not mean the players are priests.

2. E^4: Energy, Entropy, Economics, Environment

"The Earth is not a system in the sense of a political, philosophical, or mathematical system. It is much closer in kind to digestive systems, heating systems and plumbing systems."

H. Holland and U. Petersen

Preview

Part 1 provided backgrounds on Mother Nature's Two Laws (MN2L) and the quantities they govern, energy and entropy. The discussion there focused on basic thermodynamics; however, work was used to describe a spectrum of functions considerably wider than that used in introductory courses on thermodynamics. The approach in this chapter is on global scale issues. In this spirit, pollution is regarded as the mass that carries away entropy produced by the conversion of energy to work. Hence, CO_2 is just one of several entropy-containing by-products of the conversion of the chemical energy in hydrocarbons to work. It makes no difference whether the work is performed by animals or by engines typical of modern life.

One purpose of this chapter is to show readers how even rudimentary knowledge of MN2L can be used to assess current developments that affect us all: access to energy, the inevitable pollution generated when energy is converted to work, what this does to the environment, and how it affects our wallets. The position taken in this book is that scientific and technological developments coupled with economic globalization make our lives more, not less, susceptible to environmental issues. Hence, transformations of energy to work on the other side of the Earth impact our lives. However, with properly directed research, political and economic policies that are not inconsistent with basic scientific principles, and luck, consequent risks can be minimized but never eliminated.

This chapter first considers the availability and utility of various energy sources. Then attention is turned to the inevitable generation of entropy-containing products resulting from the conversion of energy to work. An appeal is made to Prigogine's theorem introduced in Chapter 3 of Part 1 as the theoretical template for minimizing the adverse effects of entropy generation. Some ramifications of entropy production for the global economy are discussed next. Then the important work of Robert Frosch on the economic benefits of recycling industrial wastes or entropy products is discussed and connected to Prigogine's theorem. The chapter concludes with a summary of the main points.

Energy

Energy gauges

Table 2.2 of Part 1 provided a gauge for the different types of energy in one kilogram of matter. For example, one kilogram of coal can release 10 joules if dropped one meter. But when burned, the same lump of coal will release 3.7×10^7 joules. In this chapter a global perspective on energy is taken. Table 2.1 below gives global energy gauges for some natural sources as well as gauges for the use of energy by humankind.

Source	Amount of Energy (joules or power)
Sun	1.1×10^{35} J/Year
Solar energy input to Earth	5.5×10^{24} J/Year
Earthquakes	5×10^{17} to 10^{20} J/Year
Lightning stroke	3.5×10^{8} J
Oceanic circulation	10^{19} J
Energy used by humankind	3.5×10^{20} J/Year
Energy from hydrocarbons	3×10^{20} J/Year

Table 2.1 Global Energy/Power Gauge

This table illustrates several points. First, there is an enormous amount of energy radiated by the sun. Earth and all the other planets receive only a tiny fraction; most of the sun's yearly energy output of 10^{35} joules is radiated into deep space. Of the total amount of solar energy the Earth intercepts, approximately 30% is reflected back into space and is not used. The net amount of energy from the sun is about 3.9×10^{24} joules/year. This energy sustains virtually all life on the planet since it is the energy source for photosynthesis. It is critical for maintaining the temperature of the planet in a range that sustains an environment conducive to life as we know it. As it also powers the atmospheric and oceanic circulations, it is responsible for weather.

The amount of energy released in earthquakes in one year is at least four orders of magnitude less than the net amount received from the sun. The energy source of earthquakes is the internal energy in the Earth's mantle. Every day earthquakes occur somewhere on Earth. Most are small and do inconsequential damage. Nearly all of the energy released by earthquakes in one year comes from one or two very large earthquakes.

The amount of energy in lightning is often overlooked; a single stroke may have more than 10^8 joules. The total amount of energy released in lightning over the course of a year is about the same as that released in earthquakes. The energy gauge for the oceanic circulation was obtained from *Atmosphere-Ocean Dynamics* by Adrian Gill, Academic Press, 1982, 662p. This number is really the available potential energy. The available potential energy for the large-scale circulation of the oceans is at least an order of magnitude larger than the kinetic energy of the currents.

The last two entries in the table were taken from *Annual Energy Review* 1993. These data show that over 86% of the energy converted to work uses coal, oil, or natural gas. The amount of energy we now use is four orders of magnitude less than that supplied by the sun and perhaps two orders of magnitude greater than natural energetic events. It wasn't until the 20th century that humankind's use of energy exceeded that in earthquakes and lightning.

To put the numbers in this table in human perspective, Table 2.2 gives some energy gauges for processes to which the reader will directly relate. These two tables illustrate the point that our environment is awash with energy. The sun has been supplying prodigious amounts of energy in the form of photons for perhaps 6×10^8 years. A tiny portion of this has been stored as chemical energy in fossil hydrocarbon compounds, which today are our chief source of energy. Thanks to the Industrial Revolution, energy is now cheap. Using a 1998 price of crude oil as a benchmark, the energy to sustain the life of a typical adult for one year costs about $6. This is only slightly more than the hourly minimum wage as of 1998. At $2 a gallon for gasoline, we in the USA pay about 50 cents more per gallon for milk. Perrier costs about three times as much, and cheap brands of beer and wine cost around five times as much per gallon as gasoline. Clearly the cost of energy for the average citizen is far cheaper than it was before the Industrial Revolution. That there is so much grumbling when gasoline prices increase demonstrates how critical energy is to the health of civilization.

Process	Amount
Heartbeat	1 J/Beat
Personal computer	5.2×10^4 J/Day or 1.9×10^7 J/Year
Consumed by an adult	10^7 J/Day or 3.65×10^9 J/Year
Consumed by humankind	1.8×10^{19} J/Year

Table 2.2 Energy Related Facts

Energy crisis: fiction or fact?

As the Industrial Revolution matured, concerns have been raised about depletion of fossil hydrocarbon fuels. The so-called "Hubbert pimple" (Figure 2.1) cleverly encapsulates these concerns. There seems to be little doubt that we are using hydrocarbon-based fossil fuels faster than Mother Nature is producing them. Is humankind dangerously close to a crisis point?

Views on this question range from extremely negative to mildly optimistic. Those from the negative camp point to oil embargoes, long gas lines and heating oil shortages in the last couple of decades, declining domestic production of coal, oil, and gas, and rising costs of fuels as evidence for a steep near-term decline in hydrocarbon-based energy sources. The general picture painted by this group is alarming. The view seems to be that humankind has passed the peak in the Hubbert pimple and that the consequences of overdosing on fossil fuels will be felt early in this century. The mindset appears to be that sources for oil and gas are like the gasoline in the fuel tanks of cars. There really is a last drop, and when it is combusted, the engine abruptly shuts down. A better analogy might be a sponge. Squeeze a little harder and more oozes out.

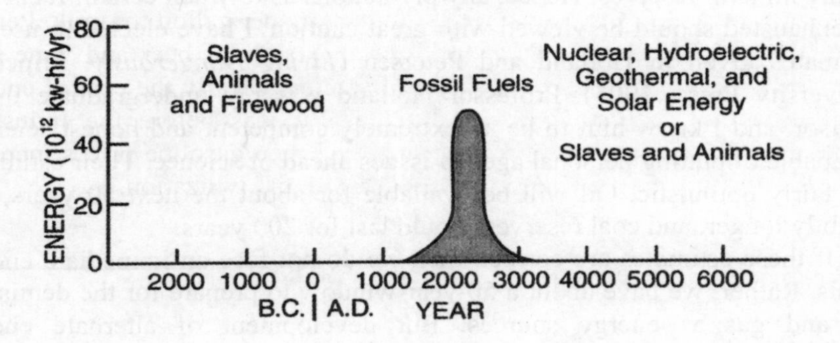

Fig. 2.1 "The energy sources of humankind, illustrating the brevity of the fossil-fuel epoch. (After Hubbert 1967)." This figure is taken from *Living Dangerously* by H. D. Holland and U. Peterson, Princeton University Press, 1995.

Geopolitical issues, national energy policies, and inflation all have significant short-term impacts on the cost and availability of energy. These factors, of course, are not directly related to the precipitous decline in hydrocarbon-based fuels envisioned by Hubbert in Figure 2.1. The pessimistic scenarios also often confuse the technical definitions of reserves and resources. According to the *Glossary of Geology*, 2nd edition, reserves are defined as the mineral or fuel deposits that can be extracted profitably under existing economic conditions using state of the art extractive technologies. Resources are defined as the reserves plus all other deposits that may eventually become available. This includes known deposits that are not economically or technologically recoverable under current conditions. Clearly, hydrocarbon resources are finite although the complete inventory is unknown.

Estimates of reserves are elastic. Late-breaking economic and technological developments can alter drastically these estimates with no change in production or consumption. Government energy policy also can play a significant role in reserve estimates. Simply mandating an increase in the cost of specific fuels with the increment passed directly to the producers would increase reserves if the increment were used to extract hydrocarbons from previously unprofitable fields. Conversely, any policy action that reduces the economic margin for producers will reduce reserves.

Clearly, there are many uncertainties in predicting the time to exhaustion of any mineral reserves. Hence, any predictions as to when certain fuels will be exhausted should be viewed with great caution. I have elected to use the estimates given in Holland and Petersen (*Living Dangerously*, Princeton University Press, 1995). Professor Holland was my undergraduate thesis advisor, and I know him to be an extremely competent and honest scientist incapable of putting personal agenda issues ahead of science. Their estimates are fairly optimistic. Oil will be available for about the next 50 years, gas slightly longer, and coal reserves should last for 200 years.

If these estimates are correct, then we do not face an immediate energy crisis. Rather, we have about a 50-year window to prepare for the demise of oil and gas as energy sources. But development of alternate energy technologies and infrastructures, even if we know what they are, takes decades and involves enormous capitalization. Holland and Petersen's estimates suggest there is adequate time to formulate a sensible energy policy that could sustain indefinitely civilization as we know it. Because of the long lead times and large capital needed to bring new energy technologies on-line, the political leadership should be developing such policies now. A 50-year time frame may be too long for energy policy to be an issue during the millennium presidential election. This is unfortunate, since any solutions likely will require decades to implement.

Desirable attributes of energy sources

Before discussing some alternatives to hydrocarbons as sources of energy, it is worthwhile to determine some desirable attributes of such sources. Four come to mind. First is efficiency. The lesson from history is that humankind has preferred efficient energy sources. Essentially we use energy transformations and fuels that provide the most energy from a given amount of mass. By converting the chemical energy in food to work an adult can release about $3x10^7$ joules in a workday. From Table 2.2 of Part 1, the same amount of energy is available from burning a kilogram of coal in a few minutes.

Energy conversion on demand is the second attribute. Humankind has insisted on the capability to convert energy to work when needed, not when it happens to be available. To meet this requirement before the Industrial Revolution, it was necessary to mobilize men and animals to convert energy to work. Clearly, this was labor intensive and thus required considerable logistical preparation for large projects. With the Industrial Revolution, large and efficient engines could be turned on and off at will. Logistic preparations necessary for large work crews were replaced by the need to raise capital to procure the requisite machines and education to design and operate them.

The third attribute is transportability. Up until the Industrial Revolution, humankind had to rely on locally available energy sources, mostly food for man and beast. Often this meant locating near energy sources. However, during the Industrial Revolution, we learned how to convert the chemical energy in coal, and later other hydrocarbons, to work. Since this was a much more efficient source of energy than renewable sources, it became profitable to mine energy ores and use part of their energy to transport the ores to users. As a result, a very efficient global infrastructure has evolved to transport fuels from their native sources directly to the user. The backbone of this infrastructure is a vast fleet of land and sea carriers that transport coal, oil, and gas from the producing areas to users anywhere in the world. This is supplemented with extensive pipeline systems that transport oil and gas across continents. Finally, there are extensive electrical power grids that transmit energy at nearly the speed of light from huge generators directly to specific appliances in our homes.

The last attribute is land use. Simply stated, we do not want large chunks of real estate traditionally used for agriculture and other essential economic functions tied up in energy production. Here, hydrocarbons have an enormous advantage over many other energy sources. The ores are normally found at depth so that competition for other uses of surface areas generally are not as severe as they are with, say, hydroelectric energy. It is quite common now to see oil or gas being pumped from the same land that is growing crops or grazing livestock. The same is true for offshore regions. Some of the planet's richest fishing grounds are in the Gulf of Mexico and the North Sea. Of course, these also are areas where significant amounts of oil and gas are produced. Traditionally coal was mined underground and towns were often built adjacent to or on top of the mines. Strip-mining is a recent innovation that departs significantly from this paradigm.

Clean energy?

Earlier it was suggested that there is no immediate energy crisis. On the other hand, there is a growing public and political perception that we are in the midst of a serious environmental crisis arising from the enormous use of hydrocarbons as humankind's dominant energy source. In thermodynamic language this is an entropy crisis. Air pollution is a long time favorite manifestation of this perception. Recently, concerns have risen that the buildup of greenhouse gases in the atmosphere is causing global warming. I discuss this in the next chapter.

A principal goal of this book is to acquaint nonscientists with the intimate interplay between energy and entropy. Thus the current energy/entropy debate is a healthy development in that it brings to the public's attention the fact that conversion of energy to work and entropy production cannot be separated. In this regard, a number of proposals have been made to shift from hydrocarbons to "clean" sources of energy. The motivation for this is that these sources do not produce significant amounts of greenhouse gases. Some of the proposals border on science fiction and thus are not appropriate for consideration here. However, four energy sources currently now in use, although sparingly, have been touted as magic bullet solutions to both the energy and entropy "crises." These are hydroelectric, wind, solar, and renewable energy sources.

Reduction of the production of greenhouse gases is certainly a positive attribute of each of these sources. However, scenarios for widespread expansion of these sources to replace hydrocarbons are overly optimistic and as yet have not addressed some fundamental issues. Here is what you might ask yourself whenever these proposals are brought forward:

1. *Hydroelectric*. In 1993, hydroelectric sources accounted for about 6.5% of the global production of energy. Essentially all existing hydroelectric plants are maxed-out, so if this source were to replace hydrocarbons, then the existing network would have to be increased by more than an order of magnitude. Construction of such facilities is expensive and it will be decades before they are on-line. Furthermore, a lot of hydrocarbon-based energy is required to build these facilities. Before such a commitment of time and resources is made, it is important to address a fundamental issue. As seen in Table 2.2 of Part 1, kilogram for kilogram, the conversion of the gravitational energy in water to work is about 300 times less efficient than converting the chemical energy in oil. An awful lot of water and a huge change in elevation at the hydroelectric power plant are required to produce the same amount of energy available from a much smaller amount of hydrocarbons. This means that hydroelectric energy sources will compete with other industries as well as individuals for the use of land and water. Thus, before a major

expansion of the hydroelectric network is undertaken, some difficult decisions on the allocation of land and water resources need to be made.

2. *Wind.* On the global scale, recovering the energy in wind to do work now is inconsequential, although it is or can be an important factor in some communities. Before committing public resources to make this a major global energy source, several technical questions should be resolved. First, the wind blows when Mother Nature decides it should, not necessarily when there is a demand for energy. Widespread reliance on this source would mean commercial and social activities would be dictated by when the wind blows, unless it is possible to store the wind-generated energy for later release on demand. Two possibilities are to store the energy in giant batteries or as heat in reservoirs. The latter is land intensive and hence competitive with other beneficial uses of land. Moreover, large reservoirs would divert significant amounts of water from agriculture. Batteries could help offset land use, but then the question arises as to the wisdom of concentrating significant amounts of lead or exotic material in the environment. Finally, it would be necessary to have backup technology when wind patterns change. Reliable backup systems tend to be expensive.

3. *Solar.* A number of proposals have been made to directly convert the energy in the photons emitted by the sun to work or other forms of energy. This technology is based on the photovoltaic effect first observed by Edmond Becquerel, a contemporary of Carnot. The effect is produced when photons interact with the atoms in certain materials to dislodge some of the electrons from their outer shells. These electrons then migrate through the material and, under appropriate conditions, produce an electric current. Although known since 1839, it has been only in the last couple of decades that any practical use has been made of this effect. This is really neat technology; it has long been used in satellites as an energy source. More recently, solar cells based on this effect have been used in some parts of the country as an energy supplement for homes and some businesses. But before this source can be expanded to replace hydrocarbon energy sources, the same issues that plague wind energy will have to be resolved. Wide scale deployment of solar panels is land intensive and there is the persistent problem of energy storage for later release on demand. Another concern that has not been widely discussed is what to do with the collectors when they wear

out. These units are very hardy and contain exotic materials that could pose some environmental risks if the widespread usage envisioned by advocates comes to pass.

4. *Renewable*. The marvelous attribute of renewable energy resources is that, although they emit CO_2 when burned to produce work, most of these emissions generally are quickly recycled back to the biota. Thus, there is little net increase in greenhouse gases. However, there are two major disadvantages of widespread reliance on this energy source. As shown in Table 2.2 of Part 1, renewable fuels such as wood are not as efficient as comparable hydrocarbons. To produce the same amount of energy or work requires a greater amount of renewable fuels than conventional hydrocarbon fuels. The other disadvantage is that growing renewable fuels will use the same productive land and water resources as does agriculture. A global shift to renewable energy would require a massive reassessment of agriculture priorities.

Another alternate energy source, of course, is nuclear. When the nuclear power industry was first started in the early 1970s, it was hailed as the magic bullet energy source. As seen in Table 2.2 Part 1, it represents a substantial jump in energy efficiency over hydrocarbons. Moreover, it enjoys many of the desirable attributes of hydrocarbon energy sources. Unfortunately the industry was saddled with arrogant and inept management that had a patronizing attitude towards public relations. Plant safety issues were not adequately explained to the public, and no comprehensible plan for disposal of spent nuclear material was put forth. Alarmists were able to establish control of the public dialogue on nuclear energy policy. Consequently, this industry is now in serious decline. In the present political climate in the United States, it is unlikely that a new nuclear power plant will be approved in the next few years. Keep in mind it takes years and substantial capital to construct a nuclear power plant. The 50-year window suggested by Holland and Petersen is adequate for a reassessment of policy concerning this energy source and subsequent construction of enough plants to offset the decline in hydrocarbon production if this is deemed appropriate.

Note that with nuclear power, mankind is nearing the theoretical limit for the amount of energy that can be extracted from fuel. Table 2.2, Part 1 showed that the amount of energy released from uranium is within just three orders of magnitude less than Einstein's theoretical limit of the complete conversion of mass to energy. After nuclear energy there will be no more million-fold increases in energy available from fuel that characterized much of the Industrial Revolution.

Entropy

The total amount of mass consumed by humankind in one year is about 1.8 x 10^{12} kilograms. In contrast, the mass equivalent of the conversion of the chemical energy in hydrocarbons to work shown in Table 2.1 is about 10^{13} kilograms. This contains about 5.5 x 10^{12} kilograms of carbon. Right now, the waste products from the conversion of hydrocarbon chemical energy to work is about five times greater than the metabolic wastes from humankind. Up until the Industrial Revolution, human and animal metabolic wastes contained most of the entropy arising from humankind's conversion of energy to work. These wastes were recycled locally back into the environment and thus did not produce global effects. As noted earlier, this was a natural application of Prigogine's principle of minimum entropy production. Any imbalances were restricted to local areas and the local inhabitants had to pay the consequences.

With the Industrial Revolution and subsequent globalization of energy production and use, this picture changed forever. The entropy-carrying products of the transformation of energy to work now have a global dimension. Thus, we need a global perspective for minimum entropy production rather than the local perspectives currently taken by political leaders and the media.

A theoretical template for a global view on minimizing entropy production is Prigogine's minimum entropy principle applied to the Earth as a system rather than to local regions. With this template, the focus shifts from local or regional limitations to global recycling of emissions of greenhouse gases.

There are several scientific advantages in this approach. First, and perhaps most important, it is consistent with the efforts to minimize the buildup of greenhouse gases through the Kyoto Protocol. As most readers are aware, this is a recent and controversial international agreement to limit the buildup of greenhouse gases by imposing national limits on greenhouse gas emissions. Although there are many shortcomings to this approach, the protocol correctly identifies natural terrestrial carbon sources and sinks as a component of the carbon budget that must be accounted for in meeting emissions standards. The next chapter discusses this protocol in detail.

A scientific group, the International Geosphere-Biosphere Programme (IGBP) Terrestrial Carbon Working Group, currently is addressing technical issues of accounting for the terrestrial carbon budget. As this group has noted, it is critically important to establish a reliable full carbon budget. The accounting system prescribed by the Kyoto Protocol is merely a subset of the full budget. The Kyoto Protocol carbon budget is arbitrary and not consistent with the way Mother Nature manages that budget. Not all industrial CO_2 emissions become greenhouse gases. A significant portion is used by the biosphere for growth, as noted previously. A nice description of the activities of this group in response to the Kyoto Protocol is given in *Science*, vol. 280, 29 May 1998, pp. 1393-1394.

Recycling of industrial wastes that carry away entropy is an important new idea. Robert Frosch has analyzed many examples where different industries and governments have learned to minimize the combined entropy production to their advantage. He has coined the term "Industrial Ecology" to describe this paradigm. Unfortunately the same term also has other connotations. Figure 2.2, taken from his article in *Physics Today* (1994), is a schematic showing how this approach works. The point is that at each phase in the product life cycle there needs to be mass feedback loops so that entropy production is minimized for all phases of the life cycle rather than for each phase independently.

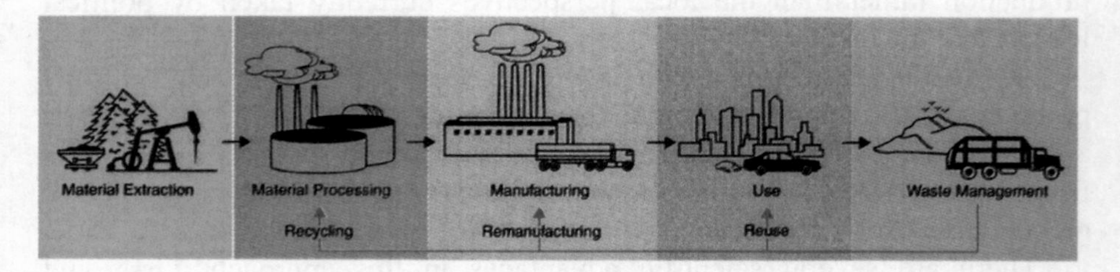

Fig. 2.2 "Product life cycle. (Courtesy of D. Navin-Chandra, Carnegie Mellon University)." This figure is taken from "Industrial Ecology: Minimizing the Impact of Industrial Waste" by R. A. Frosch, *Physics Today*, 47, 11, 63-68, 1994.

As Frosch has noted, this approach functionally emulates the way ecosystems operate. Figure 2.3 shows this in the same way Prigogine's theorem was illustrated in Figure 3.2 of Part 1. Extracted material becomes the energy source that goes to Processing. In analogy with an ecosystem, the extracted material is like solar radiation, and Material Processing is the primary producer. By consuming the output of Processing, Manufacturing and the consequent products play the role of animals. Finally, and just like a true ecosystem, there is Waste Management or the recyclers. This is where the mass loops are closed so that the entire system produces minimum entropy.

In *Scientific American* (September 1995), Frosch cites Kalundborg, Denmark, as an example where this principle has worked exceptionally well. There, an oil refinery uses waste heat for a power plant and sells sulfur removed during the refining process to a chemical company and to a wall board producer as a replacement for more expensive gypsum. Excess steam from the power plant also provides heat for aquaculture and residences.

Figure 2.4 is a schematic illustrating conditions if the refinery and power plant did not sell their waste products. Essentially, these two elements, along with the chemical company, wallboard producer, aquaculture, and residences, operate as independent open thermodynamic systems all exporting their entropy to the environment. Figure 2.5 depicts the conditions when these same elements are able to sell some of their entropy products to other elements. They then behave as subsystems of an open compound system. At this point, the potential market for waste products has a positive effect on their bottom line as well as the local environment.

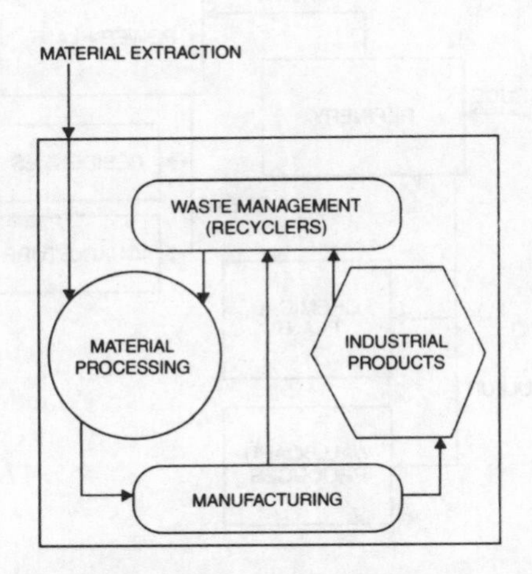

Fig. 2.3 This is a schematic for industrial mass flow. This figure is adapted from Frosch.

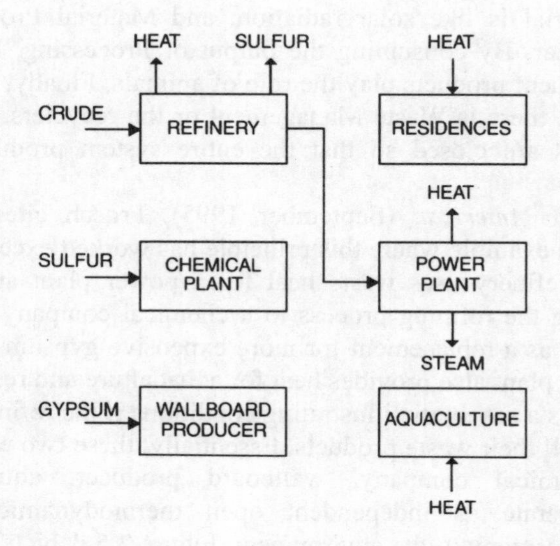

Fig. 2.4 Schematic of Kalundborg if all elements behave as independent open thermodynamic systems.

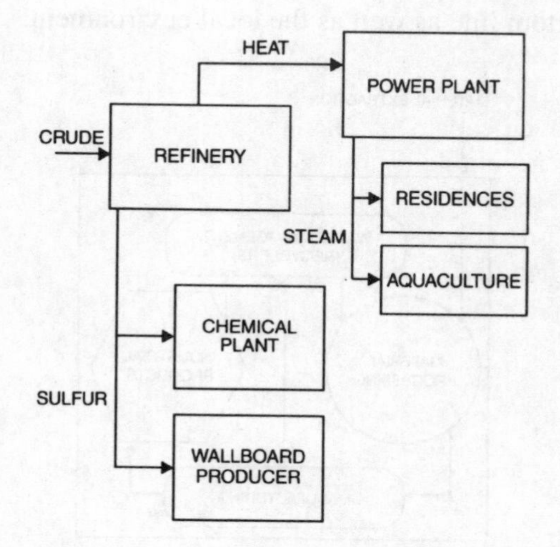

Fig. 2.5 Schematic of Kalundborg when all elements behave as an open compound system.

Economics

It may not be obvious to everyone how thermodynamics, specifically MN2L, is connected to economics. Here is the way I see it. Economics deals with the production and consumption of goods and services. The production side transforms raw materials to goods. The ethereal version of the corporal transformation is simply the transformation of energy to work. This is a First Law process. The Second Law says that this transformation cannot be perfectly efficient; there must be waste products. Moreover, the wastes must be disposed of if the economic process is to survive. The waste products appear in many forms, such as smoke discharge from factory chimneys, chemical by-products, leftover raw materials, heat, and perhaps even excess production. These wastes transport entropy out of the system. Frosch noted that industrial wastes of one industry can be raw products for another industry. Instead of being a cost, wastes could contribute to profitability.

Regrettably, the role model examples cited by Frosch are not commonplace today. Mainstream economic policy still focuses on maximizing profit by minimizing costs of production. In simple thermodynamic terms, this is a First Law issue. Such a policy ignores the Second Law, namely the consequent generation of entropy as manifested by wastes and pollution.

Ignoring the Second Law in economic processes is a lose-lose strategy in the long run. First, virtually all industrial wastes have economic significance. Finding markets for these waste products would produce new revenue sources, perhaps even causing new industries to arise. Second, if the wastes are not sold they accumulate until a crisis is perceived. Then options are limited and the bottom line is enormous clean-up costs, liability suits, and delays in production. The environment, if it is considered at all, tends to be marginalized by nebulous assumptions such as "growth is good" for the environment. As far as I have been able to determine, this has been justified by data that show rising per capita income ultimately correlates with reduction in some pollutants. In this regard, the reader may recall one of the dictums in the previous chapter: Correlations do not establish cause and effect.

The interrelations between economic growth and the environment were discussed recently by Arrow et al., (Reprinted with permission from *Science*, Arrow, et al., 268, 520. Copyright 1995 American Association for the Advancement of Science.). They point out:

> "The environmental resource base upon which all economic activity ultimately depends includes ecological systems that produce a wide variety of services. This resource base is finite. Furthermore, imprudent use of the environmental resource base may irreversibly reduce the capacity for generating material production in the future."

To put this statement in perspective, Arrow et al., 1995 cite a study by Vitousek et al., *BioScience* 36, page 368, 1986, which showed that the net terrestrial primary production of the biosphere appropriated for human consumption is about 40%! I don't have the background to assess the uncertainties in this calculation; it could be high or low. It seems obvious, however, that the percentage is much higher now than any time in the history of humankind. Our ancestors thrived even when the climate underwent much larger changes (melting of the glaciers and an enormous rise in sea level) than the worst case scenarios for global warming. One reason they were able to do so was that their economy did not require such a significant percentage of biosphere resources. This gave them plenty of options to adapt with the biosphere to drastic climate changes. Today, smaller disruptions in terrestrial primary production can ripple through the global economy.

The Frosch paradigm, which I argue is well-founded on Prigogine's minimum entropy production principle, addresses some of the concerns about environmental resource sustainability and environmental resilience raised by Arrow et al., 1997. The key point in this approach is to establish as nearly as possible steady state mass fluxes of industry entropy products. The reader should keep in mind that what is involved here is not just recycling newspapers and aluminum cans back into paper and cans. The issue is to recycle basic materials in the waste products of one industry to other industries. Unlike newspapers and aluminum cans, where the cost of recycling is more than the product, industrial recycling has to be economically sustainable.

Frosch has examined this aspect of the problem. Figure 2.6, taken from his paper in *Physics Today*, shows "Sherwood" plots for a number of industrial metals. The lower panel shows the break-even curve where it is profitable to recycle the metals rather than extract virgin ore. Under current economic conditions and policies, recycling is economically advantageous for only a few metals. These plots, of course, do not depict a static situation since both mining and recycling are energy intensive and the cost of storing metal wastes likely will continue to rise. The point is that these and other factors depict real conditions that are not adequately reflected in current economic policies. In economics, Mother Nature has a bill for entropy production, which will be paid. A realistic economic policy should minimize this bill.

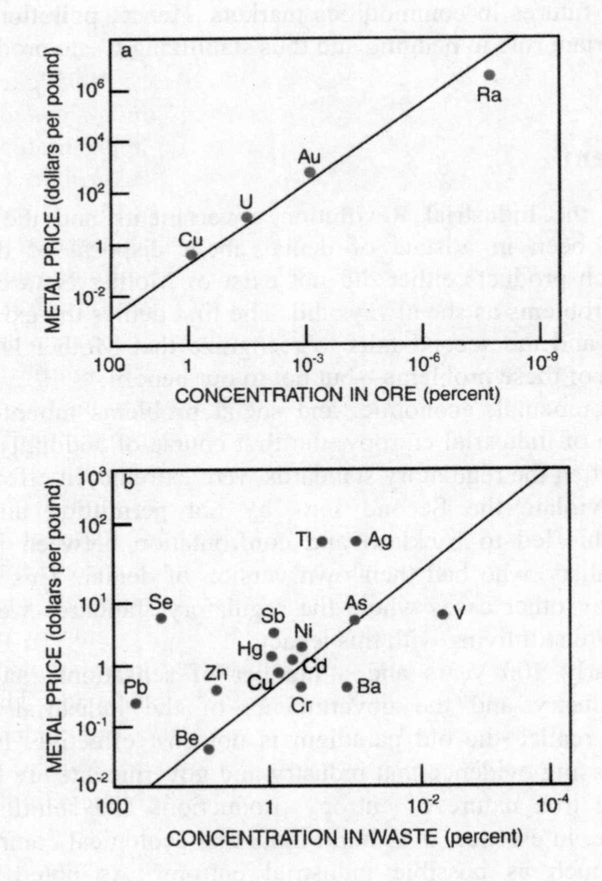

Fig. 2.6 "Sherwood plot for metals. . . .(Adapted from ref. 4)." This figure is taken from "Industrial Ecology: Minimizing the Impact of Industrial Waste" by R. A. Frosch, *Physics Today*, 47, 11, 63-68, 1994.

Much has been made by the political leadership and the media of "pollution credits" as the magic bullet solution for environmental problems. This hype ignores the fact that they do not provide any incentive to reduce pollution overall. The economic motivation is just to shift the emissions from one locality to another.

The true importance of pollution credits is that they establish economic value to industrial entropy products. Trading of pollution credits is analogous to futures in commodities markets. Hence, pollution credits can play an important role in hedging and thus stabilizing waste product markets.

Environment

For most of the Industrial Revolution governments and the captains of industry had been in a state of denial about disposal of their entropy products. Such products either did not exist or Mother Nature would take care of any problems as she always did. The first denies the existence of the Second Law and the second fails to recognize that Mother Nature indeed will take care of these problems—but not to our benefit.

As environmental, economic, and social problems mounted from the concentration of industrial entropy, the first course of action, typically, was regulatory. Often the regulatory standards were extreme, in effect mandating industry to violate the Second Law by not permitting any pollution. Predictably this led to backlash and confrontation between industry and environmentalists, who had their own version of denial. This conflict was exacerbated by other cases where the regulatory standards were not strict enough. We are still living with this legacy.

After nearly 200 years and a number of sensational and expensive cleanups, industry and the governments of the industrial nations are beginning to realize the old paradigm is not cost effective. Frosch noted there is increasing evidence that industry and governments are beginning to recognize the true nature of entropy production. The solution many are starting to use, in essence, is to build industrial ecological communities that recycle as much as possible industrial entropy. As noted earlier, this emulates Mother Nature in trying to establish global minimum entropy production by recycling the entropy-carrying products of communities of industries.

The industrial cycle of lead provides an interesting example of how an industry is attempting to minimize costs through an application of Prigogine's minimum entropy principle. Lead is an important industrial commodity, but it is a heavy metal and too much exposure is hazardous, especially to children. Thus, it serves as a useful surrogate for the general issue of particularly hazardous entropy products. Frosch (*Physics Today*, 1994) has studied the mass flow of lead in the USA and the results are summarized in Figure 2.7, taken from his work. Note that mass is reported in kilotons. One kiloton is $0.9 x 10^3$ kilograms, or for purposes here, nearly 1,000 kilograms.

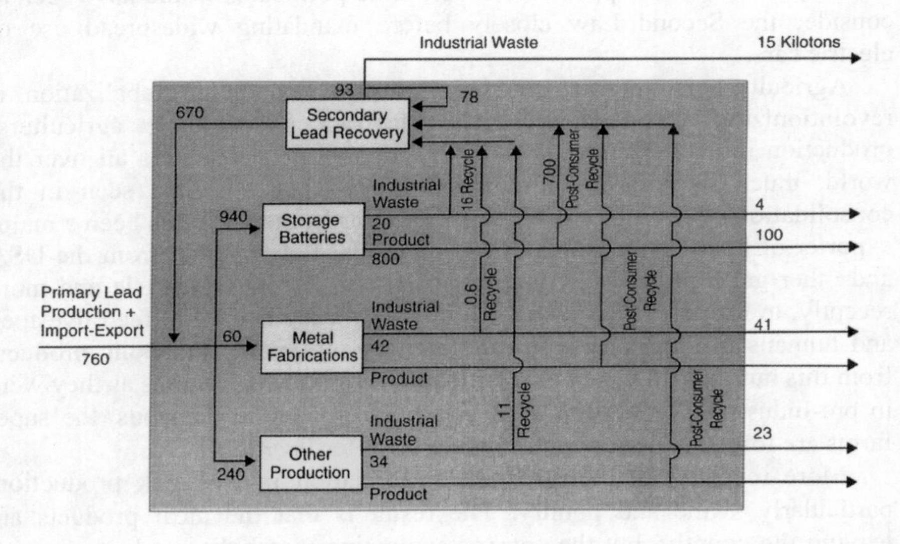

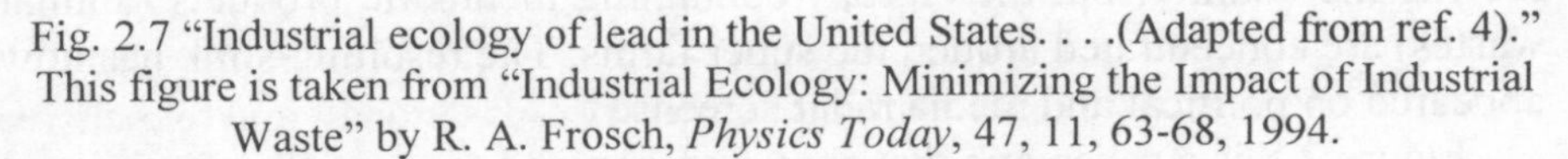

Fig. 2.7 "Industrial ecology of lead in the United States. . . .(Adapted from ref. 4)." This figure is taken from "Industrial Ecology: Minimizing the Impact of Industrial Waste" by R. A. Frosch, *Physics Today*, 47, 11, 63-68, 1994.

This figure shows that over two thirds of the industrial use of lead in the USA goes into storage batteries. Almost 90% of the lead used in these batteries currently is recycled, a very encouraging sign. Furthermore, the total amount of lead returned to the environment is just over 22% of the lead mined and imported. This indicates a substantial amount of recycling of lead back into the industry. However, as noted in the figure caption, numbers for several recycle connections were not available because the mass flows are not in steady state. Thus, this system is not yet in a state of minimum entropy production.

Lead recycling for storage batteries is in danger of breaking down. California has enacted legislation that, in effect, mandates a widespread increase in battery powered cars. This is a serious issue that is discussed in more detail in the last chapter. A vast increase in electric cars would produce a comparable increase in the demand for lead, which would produce a vast increase in primary production and import of lead. The recycle connections, which are not in steady state, cannot handle a large increase, so much of the new lead entering the system would find its way back to the environment. Clearly, a responsible approach for California politicians would have been to consider the Second Law closely before mandating widespread use of electric cars.

Agriculture is another troubled industry. Economic globalization is revolutionizing this industry. This favors concentrating agricultural production in a few "super farms" for shipment to consumers all over the world, much like fossil hydrocarbon fuels. This was first seen in the consolidation of grain farms. Since World War II, the USA has been a major exporter of grain. This means there is a global flow of food from the USA and other grain producing countries to Africa, Europe, and Asia and more recently, to former Soviet Union nations. The grain is used to feed livestock and humans in these areas. But the entropy-containing metabolic products from this nutrient flow are not returned to the producing regions as they were in pre-Industrial Revolution days. To meet production demands, the super farms are forced to use enormous amounts of fertilizers.

More recently, there has been consolidation in livestock production, particularly swine and poultry. The result is that the meat products are leaving the country, but the entropy-containing metabolic products (animal wastes) are concentrated around the super farms. The resulting stink has now appeared on political and media radar screens.

Figure 2.8 is a schematic that shows carbon and nutrient flow in the pre-Industrial Revolution. Prior to the Industrial Revolution, the consumption-entropy production connections were nearly closed and confined to local regions. Here, the decomposers were capable of recycling most of the entropy products between plants, herbivores, and humans in a nearly steady state fashion, so that each community was a minimum entropy producer. In many respects, agriculture was the archetype minimum entropy producer during this period. But in the "global village," shown in Figure 2.9, the consumption-entropy connections have been broken. Feed is transported great distances to the super farms. Moreover, the metabolic entropy products of livestock are not recycled locally back to the plants they consume. Finally, livestock is now consumed thousands of kilometers from where it was raised.

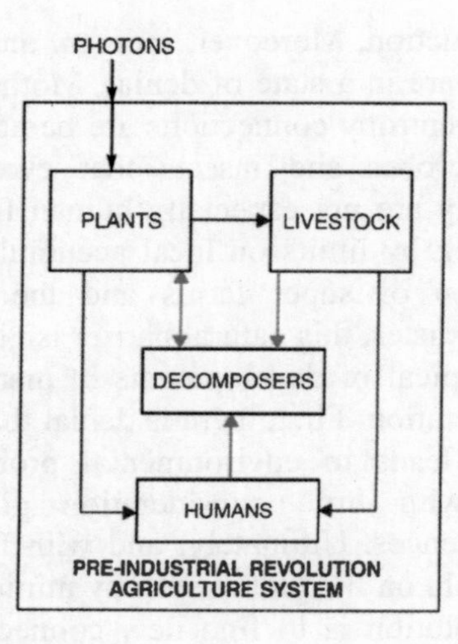

Fig. 2.8 Schematic of pre-Industrial Revolution carbon/nutrient flow. The blue arrows indicate energy flow, while the red arrows indicate entropy flow.

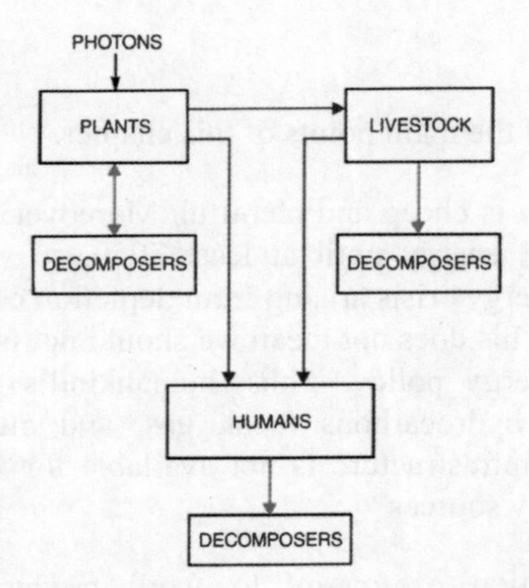

Fig. 2.9 Schematic of "global village" carbon/nutrient flow. As in Fig. 2.8, the blue and red arrows indicate energy and entropy flows, respectively.

The consumption and entropy flows in agriculture, then, are no longer in steady state; therefore, this is an industry that is no longer in a state of minimum entropy production. Moreover, industry and government agencies at virtually every level are in a state of denial. Mother Nature, of course, is not. New consumption-entropy connections are being developed that do not favor humankind. Microbes and insects that evolved to recycle these metabolic wastes locally are not especially human friendly. However, they have been kept in check by limits on local accumulations of these wastes. But with the evolution of super farms and the consequent enormous concentration of these wastes, this natural barrier is giving way.

This situation is typical of developments in many other industries that have undergone globalization. First, there is denial that any entropy products are produced. Denial leads to environmental problems, which leads to draconian solutions with little consideration given to economic or environmental consequences. Ultimately, and with luck, the industries and government may stumble on the global entropy minimization solution. With agriculture, the real solution is to find new connections for recycling the waste products so that steady state consumption and entropy mass flows are reestablished.

Summary

Here is a summary of the main points of this chapter.

1. Currently, energy is cheap and plentiful. Moreover, we are not likely to run out of fossil energy until at least 2050, so we are not likely to experience an energy crisis arising from depletion of energy resources in the near future. This does not mean we should not be developing a long-range global energy policy. Thus, humankind's dominant source of energy will be hydrocarbons (coal, gas, and oil) for decades. The technology and infrastructure is not available for widespread usage of "alternate" energy sources.

2. Any transformation of energy to work produces entropy that is transported in waste products to the environment. Because humankind is transforming more energy to work than ever before, the natural pathways for recycling entropy are becoming saturated. This is a Second Law problem.

3. This is the basis for the widespread perception of an "entropy" crisis. Examples are the general public awareness of greenhouse gases and "global pollution."

4. Most proposals, to alleviate this perception, in effect limit total carbon emissions by establishing limits on hydrocarbon fuels since there are no alternative energy sources in the short term. This, in essence, is a solution based on just the First Law. The reader might question whether this policy is achievable politically and enforceable geopolitically.

5. Some industries have started to minimize environmental degradation by forming partnerships to minimize their total entropy production by recycling waste products much like ecosystems do. This is a theoretical template that could be applied universally. Unfortunately, there is a long way to go. Currently, there are regulatory disincentives for many industries to develop markets for their waste products. Moreover, current economic policy focuses on production and ignores industrial wastes.

6. Current economic policy marginalizes the real role of the environment. Global minimization of entropy production is an important economic factor that should be included in economic policy.

Suggested Reading

H. D. Holland and U. Petersen, *Living Dangerously*, Princeton University Press, 1995. This textbook is part of the undergraduate core curriculum at Harvard. It is written for underclassmen who do not intend to major in science. It is thorough yet written at a level nonscientists can enjoy and still learn something about how the Earth works. As noted in the text, Professor Holland was my undergraduate thesis advisor. The quote at the beginning of the chapter was taken from this book.

Energy and Power, W. H. Freeman. 1971. This book is a reprint of eleven articles that appeared in the September 1971 issue of *Scientific American*. The articles are written by researchers but at a level readily accessible to nonscientists. It is a wonderful place to learn the basics of energy in the environment.

IGBP Terrestrial Carbon Working Group, "The Terrestrial Carbon Cycle: Implications for the Kyoto Protocol," *Science*, 280, 29 May, 1998. This short article nicely explains the importance of establishing a full carbon budget as well as the issues that need to be addressed.

R. A. Frosch, "The Industrial Ecology of the 21st," *Scientific American*, 178-181, September, 1995.

R. A. Frosch, "Industrial Ecology: Minimizing the Impact of Industrial Waste," *Physics Today*, 47, 11, 63-68, 1994. These two articles describe Frosch's researches in nonscientific terms. I recommend these articles as well as others cited there for all readers.

K. Arrow, B. Bolin, R. Costanza, P. Dasgupta, C. Folke, C. S. Holling, B.-O. Jansson, S. Levin, K.-G. Maler, C. Perrings, and D. Pimentel; "Economic Growth, Carrying Capacity, and the Environment," *Science*, 268, 520, 1995. This article was the result of a collaboration of scientists and economists. You don't have to be either to understand their message.

S. Hilton, *How Do They Get Rid Of It?* Westminster Press, 1970. My wife found this gem at a library book sale in Laramie. This book gives an accurate and very readable description of the end fate of industrial products. Hilton was way ahead of her time.

3. Global Warming?

"Men are never tired of hearing how far the wind carried men and women, but are bored if you give them a scientific account of it."

Henry David Thoreau

Preview

One of the most publicly debated scientific questions today is, "Is the climate getting warmer?" This question is not unlike, "Do you still beat your spouse?" Although phrased as simple yes or no questions, both are predicated on unstated assumptions or hidden agendas. Regarding the latter question, your inquisitor assumed you once beat your spouse, thus a yes or no answer is appropriate. But how do you respond if you never beat your spouse?

Many who ask the first question assume it is possible to measure the Earth's temperature to the same precision that nurses can measure a patient's temperature. Moreover, it is usually presumed that there is a reference temperature for the Earth, similar to the body temperature, and this state is "normal" for the Earth. To many, this normal temperature is a perception based on memories or tales of an idyllic past. Often there are hidden agenda issues. To some, rising temperatures are due to humankind's interference with this "natural" state and that rather simple actions can reverse the warming trend and restore normalcy to weather and climate.

This naive solution is a central focus of the public debate on climate. The scientific debate is vastly more complicated. The gist of the scientific debate is that we are starting to tinker with a primordial force of nature, and we don't know how it works.

The goal of this chapter is to introduce readers to the scientific debate. This is essential for developing responsible strategies to deal with any global climate changes. The perspective here is somewhat different from that taken by many others. The public debate has focused on the anthropogenic production of greenhouse gases and consequent temperature changes. But from Chapter 2, Part 1, the thermodynamic issue is heat, not temperature. So far, the discussion has been limited to the First Law—whether or not to limit the amount of energy transformed to work. In contrast, the view adopted here is that any anthropogenic cause of global warming, if indeed there is one, is a manifestation of the production of entropy associated with the prodigious amount of work mankind is performing. The implication of this is that any policy about global warming involves both of Mother Nature's Two Laws (MN2L), not just the First Law.

What Is Clear and What is Hazy About the Climate

Consider first the difficulty of determining the Earth's temperature. Actually there are many Earth temperatures. Every spot on the globe shows daily and annual cycles of temperature, caused by the apparent movement of the sun across the sky. This, in turn, is caused by the rotation of the Earth on its axis and the inclination of the Earth's rotation axis with respect to the plane of its rotation about the sun. At mid-latitudes, the difference between maximum summer temperatures and minimum winter temperatures is about 50 degrees Celsius or 100 degrees Fahrenheit. Superposed on the daily and annual cycles are transient weather effects. At mid-latitudes these can cause summer-like temperatures to plunge to winter-like temperatures in just a few hours. There also is a wide temperature range associated with latitude. The coldest day at the equator is much warmer than the warmest day near the poles. Moreover, there is an enormous variation of temperature with altitude. Regardless of their locations, the highest mountains are covered with glaciers year round.

Finally, there are the oceans. They cover two-thirds of the planet; moreover, the highest mountains would be submerged if they were moved to abyssal regions of the oceans. The huge volume and heat capacity of seawater means that a lot of heat energy can be stored in the oceans without a dramatic temperature increase.

The disparity between the heat capacities of seawater and land materials raises a fundamental point that is ignored in the public debate on global warming. The latter has focused on rising temperatures, but the real issue is not temperature, it is heat energy. Readers should keep in mind that global climate indices based on air and land temperatures are unreliable proxies of the heat energy changes that may actually be taking place. Global climate indices should focus on heat energy to properly take into account the oceans.

It is impossible to measure the temperature for most of the Earth's surface, so temperature observations at any one time are just a tiny sample of the Earth's temperatures. Moreover, most of these are land-based. If by the temperature of the Earth one means the average annual temperature measured at every spot, adjusted for latitude and elevation, then there will be large statistical uncertainty because of the small sample size and the great range in the numbers around the average.

In the last two decades, satellites have been used to determine the Earth's temperature. Measurements from space do not actually measure the temperature of the Earth's surface, so such measurements must be adjusted before comparisons with the ground-based temperatures can be made. With proper adjustments the latter now are in agreement with the ground-based averages. Both suggest the Earth's surface has been warming slightly since the beginning of the 1990s, but questions remain about the statistical significance of this result as well as the inferences that can be drawn from this. These temperatures do not account for heat that may be stored in the oceans. Also, keep in mind that these measurements do not establish any causes for the rise. Incidentally, reliable information on global temperatures is available at www.ncdc.noaa.gov/usextremes.html. Readers may find it useful to compare the public dialogue with the latest observations.

Recent in situ observations from the Arctic suggest more dramatic warming during this same period. There is less summer ice, and the winter ice in many places is thinner than in previous years. There are more low-pressure atmospheric events associated with higher surface temperatures, and there is intrusion of ocean water from the Atlantic, which is warmer and saltier than the ambient waters of the Arctic Ocean.

Very recently, Levitus et al. (2000) have shown that there has been substantial warming of the oceans in the last 40 years. They calculate the increased heat or internal energy of the oceans to be about 2×10^{23} joules. From Table 2.1, it is seen that this energy increase is about one-third of the annual net solar energy input to Earth. This is a substantial amount of heat, but because of the large heat capacity of water, the resulting temperature increase in the upper layer of the ocean is only 0.3 Celsius. Another curious aspect about this analysis, that likely will be lost in the public debate, is the large decadal variability in the heat content of the different ocean basins.

So, for purposes of discussion let us accept the proposition that the Earth has warmed somewhat at least since the early 1990's. This raises two other questions: Is the warming different from other interannual climate fluctuations? And what caused the warming? Some argue that the warming is the largest seen in the last 200 years. But it pales by comparison with other warming events that have occurred over longer time spans. The geologic record shows that Earth has experienced enormous climate changes just within the last 20,000 years, when the glaciers started to retreat. Some of these climate swings have been abrupt—occurring in time spans of 30 years or less! There have been significant climate changes more recently. Twelve thousand years ago the Earth experienced a dramatic cooling event know as the Younger Dryas, which lasted for a millennium. In the Medieval Age, Europe and North America warmed significantly. This was followed by a 300 year cooling event, known as the Little Ice Age, that ended about 300 years ago. Scientists do not know the causes of these climate fluctuations.

In fact, there are many possible causes of climate change. Some involve external mechanisms such as changes in the irradiance or energy output of the sun, dust clouds raised by meteor impacts, and Milankovitch cycles. The latter have been correlated with the ice ages. These cycles arise because of the long period variations in the Earth's orbital parameters. The Earth's rotation axis is inclined to the plane of rotation about the sun. In the Northern Hemisphere winter, the axis tilts away from the sun. Thus, this hemisphere receives less radiation than the Southern Hemisphere during this period. This is the reason we have seasons.

Several factors are at play in Milankovitch cycles. First, the Earth's orbit around the sun is not a circle but an ellipse. Thus, during its transit around the sun, the Earth is sometimes closer than at other times. Some readers may be surprised to learn that, currently, the Earth is closer to the sun in the Northern Hemisphere winter than it is in the summer. The variation in radiation at the maximum and minimum distances amounts to about 7% while the tilt factor is much larger. Hence, the distance from the sun is not the reason for seasons. The second factor is a very slow wobble of the Earth's axis of rotation. This wobble is just like that of a spinning top. Because of this wobble, 9,000 years ago the Earth was closer to the sun in the Northern Hemisphere summer. During this time, the Northern Hemisphere received more radiation in the summer than it does now, and correspondingly less in the winter. This factor would cause winter temperatures to be colder then than they are now, all other factors being the same. Of course, in another 9,000 years, this cycle will be completed and Earth will be further away from the sun in the Northern Hemisphere winter. There are even longer period astronomical cycles that are not considered here.

Daily, seasonal, and Milankovitch cycles are caused by easily understood mechanisms. However, they are not the only mechanisms that can produce temperature fluctuations. The most obvious, yet generally ignored in the public debate, is the sun's energy output. Models of the sun predict a general increase of energy output during the current phase of its lifetime. The scales here are very long, perhaps a 30% increase over the last 4 billion years. Superposed on this are much shorter fluctuations associated with the sun's magnetic field. The latter is correlated with sunspot activity. Observations of stars similar to the sun suggest the irradiance could vary as much as 0.7%. Recent observations of the sun show a lower value, about 0.14%. There is a wide range in periods for this variability, and there is strong evidence that even these small variations affect our climate. There are also interannual temperature cycles that have no simple explanation. The ENSO temperature oscillations in the equatorial Pacific are one dramatic and widely publicized example. The periods of this phenomena range from three to seven years.

Other mechanisms for climate change are internal to the earth. Two examples are volcanic activity and changes in the distribution of land and ocean resulting from tectonic processes. The former changes the Earth's albedo so that more sunlight is reflected back into space. The latter alters prevailing wind patterns and the oceanic circulation. In addition, seawater has an enormous capacity to store and transport heat. Thus, any process that affects the oceanic circulation can have a profound effect on climate. Recently, a third mechanism, the buildup of greenhouse gases, has gained public attention as a cause of global warming.

The Earth's climate reflects the combined effects of external and internal processes. Moreover, for chaotic systems, or ones that are hypersensitive to input parameters (recall the discussion in Chapter 3, Part 1 about Prigogine and the direction of time), even a small periodic variability in external forcing is sufficient to cause long-term, large climate oscillations. But even if the climate doesn't exhibit this sensitivity, small variations of any forcing mechanism, say solar output for example, could interact with other mechanisms to produce a pronounced climate change. A convincing example of this is given by Soon et al. (2000), who compared the sensitivity of a climate model to changes in CO_2, total solar irradiance, and just the ultraviolet portion of the solar irradiance. They found that the model climate was hypersensitive to changes in the ultraviolet solar irradiance and only mildly sensitive to changes in the other two factors. Like all careful scientists, the authors prescribe studies with different models as an additional test of the hypothesis that the Earth's climate is hypersensitive to just the ultraviolet portion of the solar irradiance.

Make no mistake about it; these natural processes are not steady in time. Moreover, they can overwhelm humankind's impact on or any attempt to fine-tune the climate. For example, a 1% change in solar irradiance over a couple of decades would produce a cataclysmic climate change. Similarly, if the Earth were shortly to enter a period of extensive volcanic activity, as it has during earlier geologic eras, the impact on the climates would be devastating. The point here is that when it comes to climate, Mother Nature holds the high cards. She can easily trump our feeble efforts to alter the climate.

Of course, we have no control over these natural processes. The only process for which there is any hope of controlling is anthropogenic production of greenhouse gases. One of the biggest scientific uncertainties is how much of the recent global warming is really anthropogenic and how much is the result of naturally occurring climate mechanisms that caused episodes such as the Younger Dryas, Medieval Warming, and the Little Ice Age. This is an active area of research, so it will take years before the scientific protocol will permit a consensus.

In the meantime, ignorance and her evil twin offspring, rumor and fear, reign. This is typified by the public discussion about the Kyoto Protocol to limit emissions of greenhouse gases. Briefly, the Protocol would require the industrialized nations to reduce CO_2 and other greenhouse gas emissions to 5% below 1990 levels by 2008-2012. As far as I can tell, the decisions regarding the time frame for the reductions and the justification for the level of reduction were politically motivated and without any serious scientific input. Before jumping to any conclusions, readers might check out the official Protocol website at www.cop3.de. This site has links to the official document and related sites.

One concern about the Protocol is that a drastic cutback in greenhouse gases translates directly to a comparable reduction in the transformation of the chemical energy in fossil fuel to work. Obviously, this has global economic implications. Some argue this will not be the case since increased efficiency would mean that the same or even more work can be achieved with less fuel. But this likely will not have much impact since increased efficiency generally means more work can be purchased for the same amount of money. Others argue that nonhydrocarbon sources can take up the slack, but, as noted in the last chapter, there are no alternative fuel sources that can be brought on-line in the Kyoto Protocol time frame.

Turn now to the buildup of greenhouse gases. How much is due to natural processes outside our control, and how much is anthropogenic? Some recent studies in the peer-reviewed literature suggest that 20 to 30% of the increase in greenhouse gases can be attributed to combustion of fossil fuels; many others are less specific. Moreover, there are major questions concerning the disposition of these gases. Perhaps a more important question is, what are the climatic consequences of continued buildup of anthropogenic greenhouse gases? If the natural processes fueling global warming continue for decades, then we should concentrate on strategies to minimize the social and economic effects of inevitable climate change rather than attempt to alter the climate. Finally, there is the question as to whether we should do anything at all and simply adapt our lifestyle and technology to accommodate the prevailing climatic conditions, as did our Stone Age ancestors.

Another uncertainty is the time frame for reducing anthropogenic greenhouse gases, assuming this is desirable. The current, apparently arbitrary, goal of the Kyoto Protocol is to make substantial reductions in a decade. Using pessimistic scenarios about the depletion of oil and gas resources, the required reductions will be achieved by fiat regardless of any international protocol. If the Holland and Petersen estimate is the lower bound on time to depletion, and if the anthropogenic estimates are too high, then the Kyoto time frame could be extended. On the other hand, if the anthropogenic estimates are too low and/or certain feedback mechanisms dominate, then we may already be in the red zone. By this scenario, the Kyoto Protocol emission targets are too low and not sufficient to reverse the warming trend. If the anthropogenic estimates are too high and/or different feedback mechanisms dominate, then efforts to reduce greenhouse gases will have a negligible effect on global warming in the Protocol time frame.

To illustrate the political issues inherent in the Protocol, consider that Canada and the USA have about 7% of the world's population, yet they produce over 20% of the anthropogenic greenhouse gases. China and India comprise about 30% of the world's population but produce only slightly more than 10% of the greenhouse gases. The standard of living in China and India is substantially less than it is in North America, and those countries have announced aggressive campaigns to improve the standard of living for their citizens. By this they mean to ramp up their capability to convert energy to work. A recent economic downturn in Southeast Asia will affect the ramp up rate, but it is unlikely to alter their basic goal. From the discussion above, their only practical alternatives are nuclear and hydrocarbon fuels. After a detailed analysis, Drennen and Erickson (*Science*, vol. 279, 6 March 1998) concluded that, even under the most optimistic scenario for development of alternate energy sources, 93% of China's energy will come from hydrocarbons, principally oil and coal, by 2025. This means that in the next two decades, China most likely will become a net importer of hydrocarbon fuels, mostly oil, since she has vast reserves of coal. It also suggests she could replace the USA as the world's number one source of anthropogenic greenhouse gases in this time frame.

The general presumption is that if the Kyoto Protocol levels for greenhouse gases are to be met, then Canada, the USA, and the other industrialized nations have some very difficult decisions to make. If China, India, and the underdeveloped countries move towards energy parity with the industrial countries, then the latter may have to make even deeper reductions to compensate for increased greenhouse gas emissions by the former. For example, the current version of the Protocol stipulates the industrialized nations are to reduce emissions on average of 5%, but the USA is to take a 7% hit. Since there does not appear to be a scientific rationale for these numbers, and because of uncertainties in the time frame and magnitude of environmental risks, there is a great political debate in the USA as to the consequences of a 7% reduction in emissions.

A 7% hit in energy use may not seem like much to many readers. One problem is that the hit cannot be applied evenly across all sectors of the economy. Considerable reduction of greenhouse gas production could be achieved by a widespread shift to public transportation. However, the infrastructure is not in place to accommodate a large change in commuting practices from private cars to public transportation. Many offices and homes now are designed to take advantage of air conditioning. Thus, an across-the-board 7% energy reduction could cause significant public health problems if this were applied to air conditioning. Agriculture is a major consumer of hydrocarbons, so a case could be made to reduce the hit for this industry. In essence, any give to one industrial or societal sector is a take from the remaining sectors. The stage is being set for a heated but not necessarily enlightening political dialogue on energy use.

These uncertainties have spawned concern by some that the present Protocol restrictions are not enough to prevent disaster, and that this is just the beginning of global restrictions on energy. Of course, a sharper reduction in the use of hydrocarbon fuels by the industrial nations could mean energy rationing and surely would cause social and economic distress worldwide as these nations would be unable to maintain present import levels with the developing nations. A point to keep in mind is that some industries would benefit by a stringent Protocol while others would be hurt. Since there is so much uncertainty, some argue that another course of action would be for the USA and its industrialized partners to void the Protocol and risk Mother Nature's potentially devastating global environmental consequences. There is also ongoing negotiation about Protocol details. These might lead to revised levels of emissions and even a change in the time frame. But, as with the present levels and time frame, it does not seem likely that they will be based on scientific analysis.

Thinking Critically About Global Warming

The public debate about global warming ignores some fundamental scientific questions. Three that seem relevant to the Kyoto Protocol are: How important is the increase in greenhouse gases *from anthropogenic sources* to climate change? Is the warming trend during the 1990s a harbinger of a major change in the climate? Will ratification of the Kyoto Protocol reverse the recent warming trend and avert a global disaster?

From the discussion above, it seems appropriate to start with the premise that current knowledge of climate dynamics is not sufficient for a clear-cut, scientifically based answer to these questions. Hence, either rejecting or ratifying the Kyoto Protocol involves some measure of risk. This is a situation in which critical thinking can provide some perspective. It is appropriate; then, to make a brief analysis of the public debate on global warming using the tenets of critical thinking as laid out in Chapter 1. In the spirit of Arons, this analysis should be regarded as just the start of a serious dialogue about these issues using critical thinking methodology. No doubt new information will alter the discussion below. Readers are invited to perform their own critical thinking analyses and to update them as more information becomes available.

1. Consciously raise questions like "What do we know? . . . What is the evidence for? . . . Why do we accept or believe? . . ." [when confronted with a problem or decision].

The previous section reviewed some evidence for recent global warming. Here we wish to examine this in more detail and to assess mechanisms that could cause global warming. For purposes of discussion, suppose global warming raises the Earth's average temperature by one degree (at this stage, temperature units such as Fahrenheit or Celsius/Kelvin are irrelevant). If this rise were superposed uniformly over the Earth, then those of us in the USA would hardly feel any discomfort. This temperature increment is at least an order of magnitude less than the daily temperature cycle for much of the country.

In fact, any global temperature change will not be spread uniformly. The one-degree global average increase could be produced by a much larger increase in the polar regions with corresponding smaller changes elsewhere. It could also be produced by a large increase in nighttime temperatures at mid and high latitudes with negligible changes in daytime temperatures. Readers may be interested to learn that both of these trends have been reported in the reviewed scientific literature. It is also quite possible that much of the heat energy associated with the greenhouse effect is stored in the ocean and is not reflected in a temperature increase on land.

Recall that there are at least three causative factors for climate change on time scales commensurate with the Kyoto Protocol. Increases in greenhouse gases and solar irradiance can cause warming; increased volcanic activity can cause cooling. It is also accepted that greenhouse gases have been increasing at least since the start of the 20th century, and that part, but hardly all, of this increase is attributed to anthropogenic sources. Moreover, all agree that an indefinite increase in greenhouse gases will cause warming. The scientific question is: Are we already in the red zone, or will the buildup have to quadruple present levels before this factor dominates other natural processes?

During this century, the land-based temperature record has not shown an increase in global temperatures comparable to the increase in greenhouse gases. This record shows fluctuations ranging over many scales. Some of the fluctuations are correlated with volcanic activity, some are correlated with solar irradiance, and others have no simple explanation. Fluctuations such as ENSO events may last for more than one year and reoccur on periods of the order of three to seven years. Without more knowledge of climate mechanisms, it is risky to infer a long-term trend based on a recent short-term increase. At this time, the possibility that in a few years the recent trend could reverse, and global temperatures could drop to levels typical of other periods of the 20th century, cannot be ruled out. Of course, no one can rule out the possibility that temperatures will continue to rise, either. Finally, keep in mind that the land and satellite temperature data do not account for any heat stored in the ocean.

The neglected mechanism in the public debate is increased solar irradiance. The research of Soon, et al., reported in *The Astrophysical Journal*, Volume 472, 1996 indicates that this factor is more important than the increase in greenhouse gases over the last 100 years. Since direct observations of solar irradiance are fairly recent, their study used proxies for irradiance. Their findings suggest some urgency in direct measurements and analysis of solar irradiance. Clearly, more analyses with longer time series of observed irradiance are necessary before the sun's effect on climate is adequately understood.

2. Be clear and explicitly aware of gaps in available information. Recognize that when a conclusion or decision is reached, it usually is done so in the absence of complete information. Be prepared to tolerate the attendant ambiguity, uncertainty, and consequences.

The gaps in knowledge are substantial. Reliable satellite temperature data are only available for the last decade, as is solar irradiance data. Thus, these two crucial data sets are not nearly as long as the data on rising greenhouse gases. The steady rise in greenhouse gases has not been matched by a corresponding rise in the surface temperatures this century. As noted above, the temperature data show substantial year-to-year fluctuations. Much, but not all, of the interannual fluctuations can be attributed to statistical uncertainty, volcanic activity, and solar irradiance variability.

In order to arrive at sound policy decisions regarding the questions raised above, it is necessary for scientists to conduct a large number of "what if" analyses. As noted above, global warming of one or two degrees in itself is not of direct concern. The real risk is in shifting weather patterns. Changing precipitation patterns are of more concern than a simple one degree increase in temperature. Some regions could become drier, others wetter, and winter precipitation at mid-latitudes could change from snow to rain and ice.

Studies of the effect on temperature and rainfall patterns, using various scenarios of anthropogenic greenhouse gas emissions, are in an early stage. Such studies, which must run the gauntlet of the scientific protocol to achieve any credibility, would be based on incomplete knowledge of the basic mechanisms affecting the greenhouse gas budgets. Recall that the scientific protocol has its own time frame. Thus, it is not likely that agreement on these studies will be reached in time to have much impact on a decision regarding the Kyoto Protocol. I stress that any policy action resulting from such studies be based on peer-reviewed results, not internal reports, books, presentations at meetings, or, God forbid, press releases or talk shows.

3. Discriminate between observation and inference, fact and conjecture.

Global warming advocates have relied on ground-based temperature observations to infer that greenhouse gases are a major cause of global warming. Global warming opponents have claimed that satellite temperatures infer that the Earth is actually cooling. There are two issues here, the reliability of the temperature observations cited by the two sides, and the veracity of the inference that greenhouse gases are responsible for any global warming. Studies published in *Science* in 1998 brought both sets of measurements into agreement. Both now indicate a very recent slight warming trend. The advocates are using this development as definitive scientific evidence of global warming. There may indeed be global warming, but this evidence is not as conclusive as advocates claim. The trend is too short to establish any long-term temperature increase. Moreover, the temperature data alone cannot support the hypothesis that the warming is due solely to the rise in anthropogenic greenhouse gas emissions.

Other observations often are cited in support of global warming. Recent examples are the particularly intense El Niño starting in 1997, the claim made by the Vice President that 1997 was the warmest year on record, and claims that huge icebergs are breaking off Antarctica. The observational record on all of these is very short compared to the greenhouse gas record and the time scales of climate changes known to have occurred since the end of the Younger Dryas about 11,000 years ago. The year 1997 may indeed have been the hottest year on record, but keep in mind, this temperature record is only about 100 years long. El Niños probably have been going on since the end of the Ice Age and long before Europeans arrived in the Americas. Thus, the size of the El Niño sample this inference is based on is 1% of the El Niños likely to have occurred. Moreover, temperature records were not as extensive or reliable at the start of the 20th century as they are now. It is quite possible that hotter years actually occurred earlier; the meteorological network was not adequate to detect them. Large icebergs occasionally break off from Antarctica simply because of the buildup of ice on the continent. In fact, evidence from deep ocean cores indicates glacial periods were periods of considerable iceberg formation. Moreover, really large icebergs will occasionally form even if there is no global warming. Without other supporting information, it is inappropriate to attribute such events to global warming. The essential point is that none of this proves or disproves global warming.

4. Recognize that words are symbols for ideas and not the ideas themselves. Be alert to situations where the same words are used as symbols for different ideas.

The global warming debate is a textbook example of this tenet. To some, "global warming" has become code for the idea that through science and technology, we are irreversibly altering our habitat. Furthermore, curtailing anthropogenic greenhouse emissions will reduce temperatures to some unspecified idyllic climate in perpetuity. To others, "global warming" is synonymous with fear mongering by tree hugging environmentalists and anti-industrialists.

As the rhetoric has ratcheted up, "global warming" has become a political litmus test. In the political environment, proponents are depicted as advocating big, intrusive government; higher taxes; and environmental policies that thwart economic growth. Opponents are characterized as proponents of big, intrusive business; higher corporate profits; and policies that put economic growth ahead of people and the environment.

5. Probe for assumptions [particularly the implicit and unarticulated assumptions] behind all lines of reasoning. [In common vernacular always be on the lookout for a hidden agenda.]

One line of reasoning seems to be that the Earth is warming because we are producing more greenhouse gases than Mother Nature can recycle. Moreover, this is reversible. Simply cut back on production of anthropogenic greenhouse gases and the climate will stabilize at some idyllic state. The other line of reasoning is that natural processes are responsible for most of the greenhouse gas buildup. Moreover, the recent warming trend is not statistically different from other warming periods. Generally, both sides ignore the role of the oceans in climate change. Changes in heat energy stored in the ocean could be more important than surface temperature.

Both sides share overly simplistic (but opposite) views about the Earth's climate. Global warming advocates take the position that we can fine-tune the climate by controlling greenhouse gas emissions, much like Alan Greenspan is said to control inflation by fine-tuning interest rates. Global warming opponents claim that the anthropogenic increase in greenhouse gases is not sufficient to cause a catastrophic climate shift. They question the direct impact of anthropogenic greenhouse gases on temperature, which they argue has no statistically significant discernable trend over the last century. In my view, neither position is substantiated by existing data.

To put some perspective on this, readers should know that during the 1970s, there was a brief cooling period when global temperatures actually fell slightly. Some even feared the start of a new ice age. Curiously, greenhouse gases were fingered as the culprit. The argument used went like this: Elevated greenhouse gases raise temperatures but also increase evaporation. Increased evaporation has a double-barreled cooling effect. It increases cloud cover, which can shield the Earth's surface from direct solar radiation, and provides more moisture to fall as snow in the polar regions. Furthermore, increased snow in the polar regions means more solar radiation would be reflected back into space and thus would not be available to heat the Earth.

The 1970s cooling scenario evoked a few carefully chosen and plausible mechanisms that could enhance cooling. However, the criticism of that scenario, as well as many of those used in the debate today, is that the relationship between greenhouse gases and the Earth's climate is oversimplified. In the 1970s, a few selected mechanisms arising from elevated greenhouse gases were used to substantiate a cooling hypothesis. Today, a few other selected mechanisms are used to substantiate a warming hypothesis. The discussion in Chapter 3 of Part 1 is relevant here. There is no simple single cause and effect relation between elevated greenhouse gases and climate. In fact, elevated greenhouse gases cause a lot of effects. Some can enhance global warming while others could promote cooling. Furthermore, greenhouse gas climate forcing will be affected by solar and volcanic forcing. In other words, it should be kept in mind that there is more than one cause for climate change and there are feedback relations between these causes as well.

Both sides use carefully selected observations and feedback mechanisms to influence the electorate. This suggests it may be worthwhile to probe for possible hidden agendas of both sides. Readers may be sure that the real issues are money and power. No action on the Protocol tends to favor an industrial status quo. A number of niche industries are pushing for the Protocol since they feel it will give them a competitive advantage. Despite some of their recent rhetoric, the large international oil producers are not likely to be affected significantly whether the Protocol is adopted or not. As noted in Chapter 2, we will be using hydrocarbon energy sources indefinitely.

6. Draw inferences from data and other information you have faith in. Recognize when firm inferences cannot be drawn.

My assessment of the data now available is that scientifically-based conclusions about how much global warming is induced by the anthropogenic buildup of greenhouse gases cannot be made at this time.

7. Given a particular situation, apply relevant knowledge, experience, and plausible constraints to visualize . . . likely outcomes that might result from . . . [possible changes or incomplete knowledge of the situation.]

The peer reviewed scientific evidence I have seen can, at best, attribute only a part of any temperature increase to anthropogenic greenhouse gas emissions. Solar irradiance and a natural buildup of greenhouse gases are two other plausible mechanisms. Then even if the anthropogenic emissions were completely eliminated, there would still be solar irradiance and natural sources of greenhouse gases to increase the temperature. So if these later two mechanisms continue, the best we can do by ratifying the Protocol is to slow an inevitable increase in global warming.

On the other hand, it is possible that slightly decreased solar activity and one or two large volcanic eruptions in the next decade could cool the planet in spite of a buildup of anthropogenic greenhouse gases. And, as pointed out earlier, it is also possible that action on the Kyoto Protocol will be too late or inadequate to prevent a climate change. It is also important to assess the consequences for the climate under a variety of actions regarding anthropogenic greenhouse gases. Scientists are just now beginning to explore these questions with climate models. Readers should expect some interesting reports on these studies in the next few years.

8. Discriminate between inductive and deductive reasoning; that is, whether the argument extrapolates from the particular to the general or interpolates from the general to the particular. [Extrapolation is riskier than interpolation but often cannot be avoided.]

The global warming advocates extrapolate from presumptions about present conditions to one possible result. This is the basis of their claim that the solution for global warming is to reduce greenhouse gas emissions by restricting the use of fossil fuels. In arriving at this, they have used inductive reasoning that one possible solution to a hypothetical special case solves the general problem. Global warming opponents also use inductive reasoning to defend their position. They argue that conditions now are the same as they have always been and natural processes can continue to handle the enormous amount of greenhouse gases produced from fossil fuels.

It gets worse. The opposing sides in the public global warming debate often corrupt their inductive reasoning with an inappropriate appeal to deductive logic. To illustrate this, I take the position of opposition to global warming; however, the logical inconsistency is the same if I were to take the position of favoring global warming. As an opponent of global warming I might claim that since the data do not clearly prove global warming, this must mean the same data proves my position. This is patently wrong. "Either or" conclusions are appropriate for deductive reasoning but not for inductive reasoning. The appropriate conclusion in this case is that the data are insufficient to prove or disprove either position.

9. Test all lines of reasoning and conclusions for internal consistency. [I find it helpful to try to apply all lines of reasoning to special cases where I believe I know what the answer should be. Be wary of lines of reasoning that do not give consistent answers.]

In time, the scientific protocol and Mother Nature will establish internal consistencies, not the media or political establishment. Longer data sets and data from different sources will be required before the scientific community will be able to reach a consensus. It may also be that by the time the scientific issues are resolved, Mother Nature will have settled the issue.

10. Stand back and recognize the logical processes being used. Are they appropriate for the situation at hand?

So far, the logic used in the public aspect of this debate does not appear to involve the scientific protocol. Essentially, both sides use selected results, much of which has not passed peer review, in an effort to hoodwink the public. Neither of the two sides has formulated any testable hypotheses. Indeed, this may not be possible in the Protocol time frame. Essentially decisions on the Protocol likely will be political.

In such a situation, the appropriate role of science is to advise policy makers and the public. The present situation is not unlike that of the 1940s when the nation's leaders had to decide whether to proceed with the production of atomic weapons. Scientists provided the technical analysis but the decision was political. We all should be concerned that in the present political environment, science likely will not play a comparable role in the decision.

11. Correlations alone do not establish cause and effect.

This is particularly true in this debate since there are at least three mechanisms that can affect global warming on time scales appropriate for the Kyoto Protocol. Correlations with just one of the mechanisms cannot possibly tell the whole story. Thus, until more information is available, it is not possible to establish whether we are really entering a period of global warming produced by anthropogenic greenhouse gases. But it is also not possible to say that we have not entered such a period.

12. Assess the consequences of accepting a false hypothesis and rejecting a true hypothesis from available data. Be aware that the attendant consequences as well as the statistical uncertainties in the two cases often are vastly different.

Take as a hypothesis that the climate is warming. The global warming could be the result of anthropogenic greenhouse gas emissions and/or natural causes. If natural causes dominate, then adoption of the Kyoto Protocol will have an inconsequential effect on global warming, but it will produce some economic consequences added to that associated with the warming. If anthropogenic emissions are the primary cause of the warming, then adoption of the Protocol might not be sufficient to ameliorate the warming trend. But if it can ameliorate the warming, then we need to judge which is worse: the economic pain of the Protocol or the warming. In other words, we do not want the cure to be worse than the disease. If the hypothesis is false, then adoption of the Protocol is obviously a costly mistake. Thus, there is risk whether or not the Protocol is adopted.

The role science can play in this debate is to assess the risks of accepting a false hypothesis or rejecting a true hypothesis and to advise the political establishment accordingly. In essence, the course of action regarding the Kyoto Protocol is a political, not scientific, decision. This means the political establishment must become sufficiently educated in the scientific protocol and be sophisticated enough to recognize that science may not supply an answer in the required time frame. Education of the political establishment will only happen when the electorate becomes scientifically literate.

The role of science and technology is not to bolster one political party at the expense of the other or to favor one group of industries at the expense of others. It is to provide wise leaders with assessments of risks concerning possible courses of action. Political decisions that impact our future need to be made with the best available, albeit incomplete, scientific and technical advice.

American Geophysical Union Statement on Climate Change and Greenhouse Gases

Both global warming advocates and opponents claim the majority of scientists support their positions. This is yet another illustration of the extremes to which these antagonists will go. So, what is the scientific view of the potential role of anthropogenic production of greenhouse gases in global warming? First of all, readers should keep in mind that this is not an issue that will be determined by a poll or an election. Mother Nature will do what she wants to do regardless of the opinions of scientists and politicians. In fact, there is no official poll of scientists regarding this. Even if there were, there are no guarantees that the majority view would prevail. Moreover, scientists are notoriously fickle. Most will alter their views to conform with the preponderence of evidence. As explained in Chapter 1, this is simply the way the scientific protocol works.

In light of this and the widespread public hyperbole, readers might be interested in the view of the American Geophysical Union (AGU) on this matter. The AGU is one of the largest scientific societies in the world. Members include scientists working on paleoclimates, scientists working on recent climates, and dynamists and modelers working on assessing the impact of greenhouse gases on climate change. In fact, virtually any scientist working on any aspect of global change is a member of the AGU.

The AGU council, which is composed of the presidents of the various sections of the Union, has prepared a well-crafted statement regarding climate change and greenhouse gases. This statement is available at www.agu.org/sci_soc/policy/climate_change.html and is appended below.

Climate Change and Greenhouse Gases

Adopted by Council December 1998

Atmospheric concentrations of carbon dioxide and other greenhouse gases have substantially increased as a consequence of fossil fuel combustion and other human activities. These elevated concentrations of greenhouse gases are predicted to persist in the atmosphere for times ranging to thousands of years. Increasing concentrations of carbon dioxide and other greenhouse gases affect the Earth-atmosphere energy balance, enhancing the natural greenhouse effect and thereby exerting a warming influence at the Earth's surface.

Although greenhouse gas concentrations and their climatic influences are projected to increase, the detailed response of the system is uncertain. Principal sources of this uncertainty are the climate system's inherent complexity and natural variability. The increase in global mean surface temperatures over the past 150 years appears to be unusual in the context of the last few centuries, but it is not clearly outside the range of climate variability of the last few thousand years. The geologic record of the more distant past provides evidence of larger climate variations associated with changes in atmospheric carbon dioxide. These changes appear to be consistent with present understanding of the radiative properties of carbon dioxide and of the influence of climate on the carbon cycle. There is no known geologic precedent for the transfer of carbon from the Earth's crust to atmospheric carbon dioxide, in quantities comparable to the burning of fossil fuels, without simultaneous changes in other parts of the carbon cycle and climate system. This close coupling between atmospheric carbon dioxide and climate suggests that a change in one would in all likelihood be accompanied by a change in the other.

Present understanding of the Earth climate system provides a compelling basis for legitimate public concern over future global- and regional-scale changes resulting from increased concentrations of greenhouse gases. These changes are predicted to include increases in global mean surface temperatures, increases in global mean rates of precipitation and evaporation, rising sea levels, and changes in the biosphere. Understanding of the fundamental processes responsible for global climate change has greatly improved over the past decade, and predictive capabilities are advancing. However, there are significant scientific uncertainties, for example, in predictions of local effects of climate change, occurrence of extreme weather events, effects of aerosols, changes in clouds, shifts in the intensity and distribution of precipitation, and changes in oceanic circulation. In view of the complexity of the Earth climate system, uncertainty in its description and in the prediction of changes will never be completely eliminated.

Because of these uncertainties, there is much public debate over the extent to which increased concentrations of greenhouse gases have caused or will cause climate change, and over potential actions to limit and/or respond to climate change. It is important that public debate take into account the extent of scientific knowledge and the uncertainties. Science cannot be the sole source of guidance on how society should respond to climate issues. Nonetheless, scientific understanding based on peer-reviewed research must be central to informed decision-making. AGU calls for an enhancement of research to improve the quantification of anthropogenic influences on climate. To this end, international programs of research are essential. AGU encourages scientists worldwide to participate in such programs and in scientific assessments and policy discussions.

> The world may already be committed to some degree of human-caused climate change, and further buildup of greenhouse gas concentrations may be expected to cause further change. Some of these changes may be beneficial and others damaging for different parts of the world. However, the rapidity and uneven geographic distribution of these changes could be very disruptive. AGU recommends the development and evaluation of strategies such as emissions reduction, carbon sequestration, and adaptation to the impacts of climate change. AGU believes that the present level of scientific uncertainty does not justify inaction in the mitigation of human-induced climate change and/or the adaptation to it.

This statement is a cogent summary of many of the points made in this chapter. Readers should refer to this when assessing any pronouncements about anthropogenic contributions to greenhouse gases.

January 2000 Status on the Global Change Issue

Political and economic strategies to deal with possible global climate change likely will be a major issue in the early part of the millennium. As more data is acquired and more detailed analyses performed, there will be a better scientific basis for action. In view of this, it may be useful for nonscientists to have a status report to start the millennium.

In the December 1999 issue of the *Bulletin of the American Meteorological Society*, Professor Tim Barnett and 10 other authors discuss an extensive effort to detect an anthropogenic component of climate change. This is the most ambitious study reported in the refereed literature to date. They conclude that near surface air temperature trends "cannot be readily explained by natural climate variability." However, warming by the buildup of greenhouse gases alone cannot explain the observed temperature patterns. The authors are careful to note the unavoidable uncertainties in their analysis.

Another conclusion, which is consistent with other recent studies, is that the recent warming trend is a combination of natural variability and anthropogenic effects. But because of the large uncertainties in the models and data, it was not possible to give a percentage for each contribution.

This is the benchmark study as of December 1999. As with all scientific results, newer analysis and data sets will alter these conclusions.

Suggested Reading

T. P. Barnett, K. Hasselmann, M. Chelliah, T. Delworth, G. Hegerl, P. Jones, E. Rusmusson, E. Roeckner, C. Ropelewski, B. Santer, and S. Tett, "Detection and Attribution of Recent Climate Change: A Status Report," *Bulletin of the American Meteorological Society*, 80(12), 2631-2659, 1999. This is the reference for the benchmark study for the start of the millennium for climate change.

J. Houghton, *Global Warming: The Complete Briefing*, Cambridge University Press, 1997. This book was written by a scientist who has been working on climate problems for many years. The first five chapters give an excellent explanation of climate processes as well as a summary of previous climate changes. Possible climate change scenarios arising from global warming are also discussed. This book is an excellent starting point for nonscientists interested in climate change.

S. Levitus, J. I. Antonov, T. P. Boyer, and C. Stephens, "Warming of the World Ocean," *Science*, Vol. 287, p. 2225, 2000. The senior author, S. Levitus, was one of my brightest students at New York University.

S. G. Philander, *Is the Temperature Rising*?, Princeton University Press, 1998. The author is one of the leading authorities on ENSO. This book is an excellent starting point for anyone interested in the basics of atmospheric processes, weather, and the difficulties of assessing global warming. This and Houghton's book are complementary. I recommend that neophytes interested in the global warming hypothesis read Philander's book before tackling Houghton's book. Philander poses a simple problem that is a wonderful parable for the scientific and technical quandaries we face at the start of the new millennium. It characterizes the concerns raised by the sealed tent analogy given in the Preface. Here is my version of Professor Philander's problem. Suppose a farmer notices a single lily pad in the pond used to water livestock and irrigate crops. The next day there are two pads, the third day there are four, etc. One day the farmer observes the pond to be completely covered with lilies and unusable for farming. What day was the pond only half covered with lily pads? The answer is the day before. Now suppose on the day the farmer noticed the single lily pad he decided to double the size of the pond in anticipation that the lilies would multiply and ultimately choke the pond. What day would the enlarged pond be full of lily pads? The answer is the day after the original pond was full.

W. H. Soon, E. S. Posmentier, and S. L. Baliunas, "Inference of Solar Irradiance Variability from Terrestrial Temperature Changes, 1990-1993: An Astrophysical Application of the Sun-Climate Connection," *The Astrophysical Journal*, 472, 891-902, December 1, 1996.

W. H. Soon, E. S. Posmentier, and S. L. Baliunas, "Climate Hypersensitivity to Solar Forcing?" *Harvard-Smithsonian Center for Astrophysics*, No. 4878, March 13, 2000. The authors of the preceding two publications are leading authorities on the sensitivity of the climate to solar forcing. I thank the authors for the preprint of the last article. One of the authors, E. Posmentier, was a colleague at New York University. We are co-authors of papers in the reviewed scientific literature.

T. M. L. Wigley, R. L. Smith, and B. D. Santer, "Anthropogenic Influence on the Autocorrelation Structure of Hemispheric-Mean Temperatures," *Science*, 282, 1676-1679, 1998. This is one of the first model studies designed to identify the role of anthropogenic greenhouse gases on global warming. The central conclusion is that both anthropogenic greenhouse gases and solar forcing are important. A number of similar studies are underway and will be reported in the reviewed scientific literature in coming months and years.

Decade-to-Century-Scale Climate Variability and Change. The National Academy Press, 1998. This is a cogent summary of the mechanisms responsible for climate change and possible societal impacts of climate change prepared by the Panel on Climate Variability on Decade-to-Century Time Scales of the National Research Council. It is written at a level accessible to nonscientists. It is a useful reference for readers who are trying to make sense of the debate on global warming.

4. Circus Earth Follies, Critical Thinking, and the Scientific Protocol

"The chief cause of problems is solutions."

Eric Severied

Preview

A moment of reflection should remind everyone how much science and technology have transformed society in just a few generations. At the end of World War II, transcontinental and transoceanic travel by train or ship took days to weeks. Now, jets routinely transport people to all parts of the Earth in less time than it takes the Earth to complete one revolution. Diseases and injuries that were permanently debilitating or even fatal in 1950 are now routinely treated in outpatient clinics. Today, in just a few hours, we communicate more information at the speed of light to anywhere on Earth than was transmitted in the entire history of the postal service. Perhaps achievements such as these have contributed to the myth that, through science and technology, humankind is conquering nature.

The thesis here is that this myth is dangerously wrong. In the Neolithic era, the consequences of natural occurrences such as earthquakes, volcanic eruptions, storms, and droughts impacted only local economies. Now such events can have global economic consequences. Solar flares or the passage of a century were irrelevant factors in hunter-gatherer and Neolithic societies. Today, the former can cause worldwide disruption of communications with considerable economic consequences. The latter was the cause of the widely publicized Y2K problem. These two simple examples illustrate that the heavy dependence of society on science and technology makes us all *more*, not less, susceptible to the natural phenomena.

Heightened vulnerability to the whims of Mother Nature is unlikely to deter our ever-increasing dependence on science and technology. Indeed, it is much more likely that esoteric scientific and technological fixes will be developed to minimize the societal and environmental risks arising from this growing dependence. Consider, for example, the Y2K problem. By some estimates, the federal, state, local governments, and the private sector may have spent more than $100 billion dollars on fixes. Simply shutting off all computers for a few minutes on December 31, 1999 was not a viable option.

One aspect for the decline of scientific literacy and critical reasoning is the ignorance, and in many cases, even disdain, exhibited by many in the media and political establishments for the way science is conducted. Media reports of scientific and technological "breakthroughs" and "predictions" often are based on preliminary or leaked results not subjected to the quality control provided by thorough peer reviews. Even worse, some news items have little or no scientific basis as would be evident if the standard of responsible journalism, independent verification of sources, was applied. Elected officials at every level of government often ignore accepted scientific practices and sound technical advice when establishing policy in favor of a perceived short-term political gain. Particularly egregious is legislation that, in effect, attempts to repeal, replace, or supersede basic scientific laws. Of course, Mother Nature has final veto authority, so we, the electorate, ultimately pay for the follies and arrogance of our leaders.

I suspect the public has a better intuitive grasp of basic scientific principles, but not necessarily the scientific protocol, than given credit for by the media and political establishments. Polls now regularly show that the public has developed a skepticism, perhaps even disgust, of the media and politics. This attitude, in part, is due to the way these establishments patronize science and technology. It also suggests that when the media and government reports on science and technology are released, there is a tendency for the public to direct their general suspicion to the reports. An alarming example of the schism that has developed between the public on the one hand, and the media and political establishments on the other, is the finding of some recent polls. In 1994, Gallop and Time-CNN polls showed that approximately 60% of adults believe in UFOs. More recently, these same polls gave an approval rating for Congress of less than half that value. The prophecy of Carl Sagan is coming true. The embrace of pseudoscientific ideas threatens the objective and critical thinking required for the existence of the Republic.

Ever-increasing reliance on science and technology, widespread ignorance of the practice of science, and broad public distrust of the media and government are key aspects of the circus earth analogy. This chapter takes a close look at a few recent circus acts performed by the media and political leaders. To set the stage for this, the scientific protocol is first reviewed, but from a different perspective than that given in Chapter 1. After this, there is a review of Mother Nature's Two Laws (MN2L).

The examples of media and government excesses deal with a variety of scientific and technical issues of considerable concern to society. Straightforward analysis of these spectacles shows that they resulted from a combination of ignorance or disregard for MN2L, the scientific protocol, and a failure to apply critical thinking. One of the issues raised by these examples, the relation between science and the judiciary, is discussed subsequently. The book closes with an envoi.

Afterword on The Scientific Protocol

The scientific protocol has evolved over the course of several centuries. It is as essential to the conduct of science as a free market is to economics. However, application of the protocol is both time consuming and frustrating for the scientists involved, as well as to those awaiting any scientific results. Unfortunately, there are no viable shortcuts. This situation is analogous to the plight of the potentate who needed an heir in one month and sadly learned that impregnating nine women did not produce the required result.

Attempts to speed up or otherwise short circuit the scientific protocol have effects not unlike those to the economy when the price of goods is artificially set by regulation and not by supply and demand. In either case, the real result is a severe correction. On the other hand, if faithfully followed, this procedure is one of the best forms of quality control ever devised. Recall is automatic and usually with no cost to the consumer. Better yet, no lawyers are involved and only the responsible parties suffer inconvenience and embarrassment.

There is another aspect of the scientific protocol that is generally misunderstood. The typical public image is of scientists working feverishly to "prove" their pet theories. This is an egregious oversimplification that is easily exposed by a moment's reflection. For every theory to be established there are one or more theories to be discredited. More often than not, new theories cannot evolve until older theories are killed off. Thus, one important goal of research is not to prove a theory, but to disprove competing or older theories. It may take only one experiment to prove a theory wrong, whereas it usually takes years of work by independent groups to assemble a convincing body of evidence for a hypothesis to be accepted as a theory. In fact, negation of a theory often is the shortcut to building a scientific reputation.

My favorite example of a "negative" result is the Michelson-Morely experiment. The latter half of the 19th century was a particularly exciting period in science, not unlike today. Technology was making huge impacts in experiments and there was an aura of impending great discovery and breakthrough. Some of the most exciting developments involved electricity and magnetism. The scientist most responsible for this was James Clerk Maxwell. His theoretical work codified all experimental results on electricity and magnetism into "Maxwell's Laws." One of the great discoveries during this period was that light propagated at a finite velocity. According to the scientific mores of those times, this implied that there was an omnipresent "ether" to support the propagation of light. The thinking was that light was fundamentally no different from sound, which cannot propagate in a vacuum.. Furthermore, so the theory went, the speed of light measured in a laboratory should reflect the motion of the laboratory relative to the ether.

Between 1886 and 1887, Albert A. Michelson and Edward W. Morely attempted to measure the motion of the Earth relative to the ether. Their approach was very clever. They measured the speed of light six months apart. When the first experiment was made, the Earth was moving in one direction, relative to the sun at about 30 kilometers per second. Six months later, when the second experiment was made, the Earth was moving 30 kilometers per second in the opposite direction. This is a large differential in velocity, which was expected to produce a clear difference in the measurements of the speed of light, assuming the ether was either at rest or moving with a constant velocity

As in all science, this experiment was not the result of a brainstorm the evening before. Great scientific and practical interest in measuring properties of the ether had been building for many years. About five years earlier (1881), Michelson had reported on some preliminary experiments in which he had not been able to detect any relative motion between the Earth and the ether. The new experiments with Morely were designed to greatly increase the accuracy of the previous experiments.

To the amazement of the experimenters, the scientific community, and the public, no motion of the Earth relative to the ether was detected. Essentially, one carefully designed and executed experiment negated a widely held assumption about the propagation of light. Although they did not set out to do this, they ended up testing the validly of the ether hypothesis.

As with all negative scientific results, the Michelson-Morely experiment had a positive effect in the long run. The failure of the ether theory was just one of a number of problems scientists had in reconciling electromagnetic theory with observations. About 18 years after the Michelson-Morely experiment, Albert Einstein proposed a radically new approach that resolved all of these problems. This, of course, was Einstein's celebrated theory of special relativity, published in 1905. It is important to note that Einstein did not set out to resolve the ether dilemma. In fact, he did not even reference the Michelson-Morely experiment in his paper, although he surely knew about it. His interest was more general: a self-consistent theory of the electrodynamics of charged bodies in motion. A by-product of his theory was the resolution of the negative results of the Michelson-Morely experiment. One of the great attributes of the scientific protocol is that it provides the means whereby the resolution of a scientific question can come from a completely new direction rather than a direct assault.

The initial reaction of many scientists to special relativity was that it was just another complicated and doomed attempt to explain the fact that the motion of an observer did not affect measurements of the speed of light. For example, in 1919, C. E. Abbot, home secretary of the National Academy of Sciences, wrote to G. E. Hale, "I pray to God that the progress of science will send relativity to some region of space beyond the fourth dimension, from whence it may never return to plague us." (This quote was reported by J. D. Fern in "The Great Debate," *American Scientist*, 83, page 410, 1995). It was not until the early 1930s that enough experimental data had been amassed to provide some consensus on verification of relativity. This stands in sharp contrast to the immediate acclaim of the Michelson-Morely experiment among scientists.

Recall there are three legs to the scientific protocol. The first is the use of the scientific method in conducting research. This is an exquisite exercise in critical thinking. It requires the investigator to shed personal biases, evaluate hypercritically one's procedures, and view all results objectively. This is extremely hard to do. The second leg of the protocol is open sharing of data. Of course, the scientists who perform the experiments or collect the observations are given reasonable times to check and analyze the results and submit papers for publication. But after publication, the data must be available to all. Perhaps I have been fortunate, but I have found other scientists pleased to share their data with me, even before publication of their papers. Of course, if they were not co-authors, I acknowledged the source of the data.

The third leg of the protocol is publication of the results in appropriate refereed journals. This is the leg nonscientists should focus on the most since it encapsulates the first two legs. Publication of scientific results, in peer reviewed journals, is as important to scientific progress as a free market is to economic growth. The old saw that a job isn't over until the paperwork is completed applies here.

For reasons I do not understand, this leg of the scientific protocol has come under attack in recent years. There seem to be two themes to these attacks. One is that refereed publications are simply a way for established scientists to suppress dissent and maintain the status quo. This is patently wrong. Refereed publications do not suppress dissent; they encourage it! The other theme is that the Internet and the need for rapid dissemination of the latest breakthroughs make refereed publications obsolete. This may play well with some futurologists but it does not reflect what is really happening. By any sensible measure, the peer review process has adapted quite well to technological developments in publishing. Most primary journals encourage email reviews and many rapid-publication or letters journals accept papers submitted on-line. These papers are posted on secure sites that can only be accessed by designated reviewers. All of this is speeding up the review process and maintaining necessary quality control. In fact, peer review is one of the very few activities on the Internet where there is some measure of quality control. In view of this, one must be suspicious of nefarious attempts to "fix" a procedure that is functioning properly.

Many in the media and political arenas tend to view drafts and preliminary reports as "smoking guns" or "deep throat" sources and thus worthy of broad exposure. This may or may not be true for nonscientific matters, but in science such a view is dangerously wrong. Readers should be aware that scientists who provide media access to unreviewed scientific results know that these actions seriously jeopardize publication in the reviewed literature. Applying one of the tenets of critical thinking to this situation suggests these individuals may have hidden agendas. Rather than blindly publicizing unreviewed scientific results, the media would do better to investigate the agenda. The most essential, but by no means only, requirement for any scientific result to have merit or value is to pass peer review. Drafts of papers or technical reports do not fulfill any scientifically based value criteria. There are many cases where authors have changed their minds about critical points of papers before journal publication. Such changes often are the result of valid criticisms arising from peer review.

It must also be kept in mind that publication in a refereed journal is a necessary but not sufficient condition for authenticity of scientific results. The analogy in business is that bringing a product to the marketplace is not the same as earning a profit on the product. Almost every issue of any scientific journal carries corrections to previous papers and sometimes very lively debate on the accuracy of prior publications. These debates, which may last for years, are essential aspects of the scientific protocol. Assessment of published results in principle never ceases. It follows that if scientists are skeptical of published results until there is corroboration, then nonscientists should completely ignore all results that have not passed peer review.

To the outsider, the scientific protocol may look like a raucous melee of quarrelsome scientists, perhaps not unlike the floor of the New York Stock Exchange (NYSE) as viewed by tourists. Being first to disprove a hypothesis or to report on a new phenomenon is not unlike buying low on the Exchange. One difference is that at 4:00 p.m., the bell rings at the NYSE, and trading on the floor ceases until the next day. Scientific research has no geographic boundaries or time constraints.

Afterword on Energy and Entropy

As I have tried to show, transforming energy to work makes life and economics possible. The essence of the Second Law is that not all the energy stored in fuel is available to do work. The measure of the energy unavailable for work is entropy, which must be exported from the system in waste products. Moreover, unless nuclear processes are involved, the total amount of mass in the waste products must equal that in the fuel. The limiting case is a closed system, which does not interact with its environment. But even here, the Second Law states that its entropy must increase with time.

An analogy may be appropriate at this point. Imagine the reaction of members of a dinner party at a four-star restaurant when the waiter presents the bill. Some guests go through a period of denial about the existence of the bill, presumably in the hope that someone else will pick up the tab. Others may argue about what they perceive to be their fair share. Since the start of the Industrial Revolution, and particularly since World War II, the USA has been gorging on energy. Mother Nature's bill for this feast is entropy. This bill is as certain as death and much more certain than taxes. Accepting its inevitability, doesn't it make sense to minimize the bill while simultaneously extracting as much work as possible from fuel? As participants in the circus, we are behaving just like the dinner party. Some are in denial about the existence of Mother Nature's entropy bill; they regard the feast as Mother Nature's treat. Some feel the bill is the sole responsibility of the polluters, while others feel it is the government's responsibility. What is lacking in the current public hubris is a scientifically based debate about the types of energy to transform and how to divide the costs of the entropy bill.

It is virtually a certainty that humankind will continue to increase the amount of energy converted to work in decades to come. Furthermore, most of the fuel will come from fossil sources for the foreseeable future since appropriate technology and infrastructure to produce and use other fuels simply is not in place. It is both costly and time consuming to develop the infrastructure to convert other fuels to work. We have almost 200 years experience tapping into the chemical energy of fossil fuels. What has not been addressed yet is how to manage the consequent entropy production for our own good.

It is emphasized that MN2L apply to all industrial products. The production of goods always involves transformation of energy with an attendant production of entropy. This rather obvious fact needs to be acknowledged in effective energy and industrial policies. Prigogine's minimum entropy production principle, as embodied in Frosch's Industrial Ecology concept, is an excellent starting point for a scientifically based energy/entropy policy.

Economic globalization is currently in vogue although it really started with the Industrial Revolution two centuries ago. It is appropriate, then, to take a global view of entropy production. In this context, the reader should keep in mind that the Earth, to a very good approximation, is a gigantic closed thermodynamic system, or sealed circus tent, at least on social and political time scales. All its component subsystems must be in reasonable balance in their production of entropy if the gigantic closed system is to continue to operate under conditions favorable for humankind.

Public debate on energy and entropy policies will continue indefinitely, so readers should stay alert and be suspicious of new pronouncements and developments. Here are some particular issues to watch out for. In regard to the First Law:

1. *Any process, no matter how simple, requires energy.* This means that fuel must be transported some distance to sites where it is used in the energy transformation to work. Something as common as plugging into a wall outlet requires the flow of electrons to perform desired functions, or work. Of course, a power station has to be at the other end of the circuit to pump the electrons through the power lines to the wall outlet. The power station must get its energy from other locations. The most common forms of matter used in the transformation of energy are hydrocarbons. In certain regions, energy can be obtained more or less directly from sunlight, wind, and ocean tides. These sources can provide useful energy supplements in these areas, but they should not be counted on as replacements for the global dependence on hydrocarbon based fossil fuels.

2. *Energy is not only equivalent to mass; it is also equivalent to money.* Recall from Chapter 2 of Part 1 the type of energy transformation determined the amount of energy available from a unit of mass. Similarly, there are different money equivalents for each of these transformations. The law of supply and demand determines these money equivalents. They also depend on location, economic conditions, and government policies.

In regard to the Second Law:

3. *The transformation of energy to work will always produce entropy that is contained in waste or poisonous by-products.* It is essential that these by-products be removed from the system if the system is to continue to function, as these wastes generally are harmful. Proper disposal of this waste requires expenditure of more energy/money. Prigogine's theorem for minimum global entropy production indicates that efficiencies can be

achieved when waste by-products from one industry are used by other industries in sufficient amounts that an approximate steady state flow of by-products results. However, it is important to acknowledge that this produces only a state of minimum entropy production; zero entropy production will never be achieved.

4. *The by-products are mass, which has an energy/money equivalent. Thus waste by-products have economic implications whether or not they have other industrial uses.* They could be treated as commodities, in which case markets can develop. If the by-products have no industrial economic value, they must be removed from the system and stored elsewhere. In effect, they become unusable real estate, which also has economic and tax implications. In the latter case, there is also a cost of transporting the by-products to their storage sites and maintaining security at these sites. The inability of political and media leaders to recognize the economic aspects of by-products of controlled energy transformations in particular, and the Second Law in general, is a great failure of our version of capitalism.

5. *A sound energy policy must address consequent entropy production*. In essence, MN2L must be used simultaneously, not individually.

6. *A comprehensible industrial energy/entropy policy must include plans for what to do with industrial products after their useful life.* Examples where such policies are needed range from the widespread proliferation of plastics to disposal of spent nuclear materials and hazardous wastes. The lack of any policy regarding this issue is one of the clearest indications that the media and political leaders are in denial about production of entropy.

This item also touches on another aspect of the Second Law that is generally ignored by the media, business, and political establishments. This is the sequestration of wastes containing harmful entropy products. The theory behind sequestration is that the wastes can be encased in an isolated thermodynamic system. There are two issues here. First, as noted in Part 1, it simply is not possible to construct an isolated thermodynamic system that can remain isolated indefinitely. The environment may change causing containers to leak or rupture. But even if the system remains isolated, its entropy will still increase. The molecular structure of containers ultimately breaks down or, in thermodynamic terms, it evolves to a state of higher entropy. This evolution will be exacerbated by chemical interactions with both the wastes and the environment and by nuclear reactions with wastes containing radioactivity. The longer the predicted time to breach the greater the uncertainty of the prediction.

The same reasoning applies to synthetic implants in the human body. The implants cannot behave as isolated thermodynamic systems indefinitely. Ultimately, they will breach or break. This possibility and its consequences should be a factor in any implant risk assessment, as will be discussed in the next section.

This leads to the last item regarding the Second Law.

7. *No system can be isolated from its environment indefinitely.*

In the next two sections, four examples of media and political misrepresentation of recent developments involving science and technology are summarized. These examples created public concern as well as a drain on public coffers. In the critique of these examples, I try to show readers how application of critical thinking, knowledge of the scientific protocol, and an intuitive sense of MN2L would have exposed the hyperbole.

Media Follies

The New York Times and the Yucca Flats Flap

Synopsis

The 5 March 1995 issue of *The New York Times* carried an article with the provocative title, "Theory on Threat of Blast at Nuclear Waste Site Gains Support." The material for the article, an internal report from the Los Alamos National Laboratory, claimed that a nuclear waste repository site planned for Yucca Mountain, Nevada was unsafe and, under certain conditions, leaking wastes could concentrate enough radioactive material to produce a nuclear explosion. That article set off its own chain reaction. The congressional delegation from Nevada understandably was up in arms, as were state government officials and environmental watchdog groups.

However, the report was not accurate. The scenario depicted in the report had been looked at before by other groups at other laboratories and found not to be credible. Nevertheless, the widespread publicity and political volatility of the subject matter forced the Department of Energy to hire outside consultants to reassess the earlier conclusions. Their findings supported earlier studies that concluded the hypothetical threat suggested by the report was not credible. This costly study took about a year to complete.

This was not the first time *The New York Times* had stirred up controversy. The 18 November 1990 issue published the article, "A Mountain of Trouble." This was based on a technical report by a different researcher who contended that the Yucca Mountain site previously experienced periods where the groundwater had repeatedly immersed rocks that were now well above the groundwater table. Even though that report had drawn substantial criticism from other scientists, the Department of Energy spent $300,000 to have it debunked by a special National Research Council study.

It is tempting to finger *The New York Times* reporter who wrote these two articles as the culprit. This assessment, however, avoids the real issue. There was nothing in either the internal or technical reports that suggested an immediate crisis justifying immediate public disclosure. Presumably, the decision to publish these articles before conducting a thorough investigation, including verification by independent sources, was editorial policy. It is stressed that this type of policy deficiency regarding science and technology is not limited to *The New York Times*. I have examples of similar distortions from many major newspapers and the national commercial and public television systems.

Critique

Was there a tip-off discerning readers might have used to ignore *The New York Times* articles? Yes. The articles were not based on any research that had appeared in a peer reviewed scientific journal. In defense, *The New York Times* claimed it used an assessment by another scientist as independent verification. Independent verification is a cornerstone policy of responsible journalism. However, *The New York Times* implementation of this policy was based on experience in nonscientific arenas. In science, the minimal independent verification is peer-reviewed articles. Thus, the message to readers is clear. Noteworthy scientific and technical news that is fit to print should be based on reports that have run the complete gamut of the scientific protocol.

Disposal of spent nuclear material is an important societal issue. The technical issues are among the most complicated mankind has ever faced and there is considerable uncertainty about many critical details. In fact, articles on nuclear waste disposal at Yucca Flats and other prospective repositories appear regularly in refereed scientific journals such as *Science*.

The only arena where these issues have any chance of being resolved in a rational fashion is the reviewed scientific literature. Here there can be an open and frank discussion on the analyses with the same data available to all for independent assessments. The proper role of the media is to report aspects of the scientific and technical debates and to summarize findings that are based on the scientific protocol, not sensational claims in technical reports.

The development of the nuclear industry is also an example of the long-term havoc that can result when policy makers ignore the Second Law. Here, mass is transformed directly to energy and ultimately to work. The perspective taken here is that the waste products of the transformation contain that ethereal quantity, entropy. The amount of waste material and the nature of the radioactivity can be calculated quite accurately before any weapons are made or nuclear plant built. Nevertheless, policy makers chose to ignore the consequences of improper waste disposal. If the waste issue had been addressed up front, then there would have been no Yucca Flats controversy.

Essentially two solutions are available: sequestration or recycling in the spirit of Frosch. The first solution is based on the premise that the waste containers will behave as isolated thermodynamic systems for very long times. As noted earlier, this is risky, but at this stage, it may be the only option left. Currently there is considerable discussion in the reviewed scientific literature dealing with this factor. Regarding the latter solution, the radioactivity emitted by the wastes constitutes another face of energy. With forethought some industrial uses for the wastes might have been found. However, because of the large buildup of radioactive wastes, it will be difficult to come up with a sensible recycling scenario within an acceptable time.

Media Quake at New Madrid

Synopsis

In the late 1980s, a business consultant, Iben Browning, predicted a major earthquake would occur at New Madrid, MO on 3 December 1990. Browning had an interesting background. He had an undergraduate degree in math and physics, and an M.A. and Ph.D. in zoology. He published four books on such diverse topics as the impact of climate on man, robots, and AIDS. He also held over 60 patents. He wrote a newsletter with a devoted following and he was a consultant to Paine Webber and a number of other businesses. These are all noteworthy credentials for one in the advisory business. Conspicuously absent from his vitae was a list of refereed scientific publications in any field.

Browning had claimed successful earthquake predictions before the New Madrid episode; however, they do not stand up to a posteriori assessment. They were either vague (major earthquakes will occur in the next six months in earthquake zones) or otherwise nonspecific. His methodology was based on identifying periods of higher than normal tidal loadings from special alignments of the sun and moon and the planets. This approach was tested rigorously in seismology before Browning and was shown to have no predictive capability. The essence of the criticisms of this approach is that these maximums are inconsequently larger than the normal tidal loadings and substantially smaller than typical tectonic loadings.

In a talk at a business seminar in Atlanta in February 1989, Browning stated that a serious earthquake *might* strike near the Memphis area in early December the next year. This "prediction" circulated by word of mouth until late November 1989 when the AP and the *Memphis Commercial Appeal* picked it up. In December, at the Missouri Governor's Conference on Agriculture, Browning predicted a major earthquake in the Memphis area on 3 December 1990. Two days before this talk, he told a special breakfast group of clients about the earthquake threat and also stated, "there is a 50-50 probability that the federal government of the United States will fall in 1992."

It is natural that anything related to earthquakes would strike a chord in this region. An active seismic zone cuts across this part of the country, and one of the most severe earthquake series ever to strike North America took place between late 1811 and early 1812 in the New Madrid area. Although there was no scientific basis for Browning's predictions, the print media from the "Show Me" state as well as adjacent states followed this up with a series of alarmist articles. By the summer of 1990, local governments and much of the population in this area were in a state of panic. Emergency response preparations were underway, plans were being made to close schools and businesses in December, and sales of earthquake insurance riders were at record levels. By the fall this had become a national media circus. "Good Morning America" interviewed Browning, and special segments about the New Madrid earthquake were aired on the "Today Show," ABC News, and "NOVA" on PBS. NBC ran a mini-series on earthquakes in Los Angeles, and *USA Today*, *The Wall Street Journal*, and most other "mainstream" newspapers carried articles on Browning's earthquake prediction. The general public hysteria caused many children in the area to have psychological problems requiring counseling.

I had an educated and intelligent elderly aunt, who lived in Murray, KY, which is in the western part of the state, not far from New Madrid. I visited her in 1990 and, although I am not a seismologist, she quizzed me closely about the 1811-1812 earthquakes, solid earth tidal loads, the status of scientific earthquake predictions, and the consequence of another earthquake of magnitude similar to the 1811-1812 earthquakes in this area. I explained to her that the tidal loading in early December would be no larger than other peak times in the last 30 years and that they had not triggered any earthquakes. Furthermore, as far as I knew, the tidal loading theory had no scientific basis. I did my best to assure her that although the New Madrid fault most likely would produce another large earthquake, predictions of when and where it would occur were no more reliable than horoscope forecasts. She seemed to be assured by this; nevertheless, it was upsetting to see this wonderful lady experience such unnecessary angst near the end of her life.

Newspaper accounts indicate that 1-3 December at New Madrid was a circus. This small town of approximately 3000 was overrun by thousands of curious tourists, media, and hucksters. Sunday, 2 December, was an unusually active day for the churches. Cars with loud speakers advised all to prepare for the end, sales of T-shirts with appropriate messages were brisk, and "quakeburger" was the popular special at a local eatery. The national networks had remote units on the scene for live coverage. News teams from Japan and Great Britain were even present. The militia of all the affected states were either mobilized or on call. Everyone came to the party except the guest of honor, fortunately.

Critique

All of this turmoil was in response to an unscientifically based earthquake prediction from an individual with neither training nor experience in seismology, nor any credible record of seismic prediction! The turmoil would never have occurred if the media had adhered to a policy of relying only on information obtained from the scientific protocol before publishing or airing their articles. Seismologists and responsible scientists, both in the region and throughout the country, repeatedly pointed out the deficiencies in both the predictions and Browning's credentials, but to no avail. The hysterical media coverage caused considerable public anguish. This, in turn, caused disruptive and costly emergency preparations in many communities throughout the region. Area businesses relying on seasonal sales during December took a huge hit.

Another particularly galling aspect of this episode was the conspicuous lack of critical thinking on the part of the media and local political leaders. Questions like "what do we know? . . . what is the evidence for? . . ." were never raised. No attempt was made to identify gaps in information and there was no discrimination between observation and inference, fact and conjecture. No effort was made to probe for implicit or unarticulated assumptions, and relevant knowledge was completely disregarded. It was never recognized that Browning's argument was based on an extrapolation that a slight increase in astronomical tidal loading would cause earthquakes in New Madrid. If this argument were true, then would not other earthquake zones be equally at risk? Finally the political leadership failed to assess the consequences of accepting the false hypothesis that an earthquake would occur.

It should be pointed out that not all the media were taken in. The U.S. Geological Survey Circular 1083, from which the above was extracted, provides a sample of the newspaper coverage during this period. Several newspaper editorials pointed out Browning's background and the suspicions regarding his predictions. My favorite was the decision by the editor of the *Paducah Sun*, Jim Paxton, not to publish any earthquake-related stories from 25 November until the weekend after 3 December 1990. I also commend his editorial of 18 November 1990 as an outstanding example of journalistic critical thinking about the earthquake threat. It should be required reading for all journalists.

Predictably, the media made a scapegoat out of Iben Browning and thus deflected the real blame from themselves. In fact, Browning merely exercised his First Amendment right while providing his clients with services for fees. Caveat Emptor! On the other hand, the media had a public responsibility in this matter. Collectively they failed in such a simple duty as checking Browning's credentials and previous predictions, while actively promoting panic in the population. Elected officials hardly covered themselves with glory in this case, either. In view of the widespread publicity and fear generated by the media reports, it is understandable why they closed schools and went to emergency alert. But this had the effect of adding to the panic.

Widespread dissemination of doomsday "predictions" by unqualified individuals through "responsible" elements of the media is no different than blatantly crying "Fire!" in a crowded building just to see the panic. In this case, the media caused considerable extra expense to local communities and angst to the citizenry through ignorance and disregard of the scientific protocol.

Government Follies

Silicone implants

Synopsis

After years of evaluation by appropriate government agencies, several large pharmaceutical corporations began promoting silicone breast implants in 1962. The procedure was approved for reconstructive surgery following mastectomies, but it was never recommended for uses such as cosmetic enlargement. Nevertheless, between 800,000 to 2 million women elected to undergo the breast implant procedure. This is considerably more than the number of mastectomies performed during this period. Almost from the beginning, there were horror stories about problems and reactions to the implants, ranging from leaky implants, to shifts in implant location, to cancer and autoimmune diseases. These difficulties resulted in a large number of suits brought against the pharmaceutical companies and the doctors who performed the procedures.

These suits were vigorously defended, and it wasn't until 1991, almost 30 years after the procedure was approved, that a litigant finally won. This was a woman who claimed that the implants had produced an autoimmune disease. She was awarded $7 million in a jury trial. A common feeling at the time was that the large pharmaceutical firms were grossly negligent and it was only fair they should pay for ruining these women's lives. In response to this outcry the FDA severely limited the use of silicone breast implants.

That trial was remarkable in the amount of technical medical testimony it produced. Even more remarkable was the fact that virtually none of the technical testimony for the litigant was based on peer-reviewed research. Subsequent peer-reviewed research by the American Cancer Society, the American Society for Clinical Oncology, the American Medical Association, the American Society of Plastic and Reconstructive Surgeons, the British Council on Medical Devices, the American College of Rheumatology, and the FDA itself found no evidence that these implants cause autoimmune diseases. Nevertheless, 400,000 women are involved in a class action suit against the major producers of the implants and an additional 20,000 to 30,000 women are involved in individual lawsuits. In 1998, a $3.2 billion dollar settlement was reached in a class action suit against the primary maker of implants, Dow Corning. However, approximately 6% of the litigants have not agreed to the settlement, and so this saga continues into the new millennium.

Critique

Of course we are all concerned about the plight of women who experienced leaky or shifting implants, and anyone who has cancer or autoimmune disease, regardless of the causes. The concern here is not whether the judgements were fair or whether mainstream medical science had been unable or negligent in detecting a real connection between the implants and autoimmune diseases. These issues now have been settled without regard to the scientific protocol. The point is judgements based on allegations that autoimmune diseases are caused by implants were not justified by the refereed scientific literature. It is possible that proper attention by the judiciary to the scientific protocol might have provided fair and just awards based on negligence by plastic surgeons or manufacturer liability for other health problems caused by leaky implants. But unfair product liability judgements such as outlined here will lead to denial of beneficial new technology or liability being underwritten by the federal government. The latter scenario can lead to gross negligence by some corporations.

A deeper concern, which this matter identified, is the appropriate role of the scientific protocol in legal matters. In most cases, the judicial system ignores the protocol and relies all too often on data that may not be available to the scientific community and analyses that have not passed peer review. In product liability litigation, both parties have teams of scientific and technical "experts" each claiming to provide "scientific" evidence in favor of their clients. Even in capital crimes, justice often seems to favor the side with the most credible and/or charismatic scientific and technical "experts." In state courts the standards for admission of technical evidence often are lax, hence charismatic witnesses or pitiful plaintiffs are able to sway juries even though the supporting science and technology is suspect. Moreover, scientific and technical "expert" testimony under current legal practices is expensive. This contributes to exorbitant settlements based on pseudo scientific and technical issues that have not passed the quality control filter of the scientific protocol. Decisions based on weak science undermine public confidence in the judicial system.

From another perspective, both the women who elected to undergo this procedure and the medical profession would have benefited from a simple qualitative analysis based on the Second Law. As noted earlier, the human body behaves as an open thermodynamic system. We take in mass in the form of oxygen and food, convert the chemical energy to work, and expel entropy in solid wastes and CO_2. Implants are also thermodynamic systems. The expectation was that when inserted in the human body, they would act as isolated thermodynamic systems and thus would not interact with the body. This simply is not consistent with the Second Law. At the molecular level there will be interactions between living cells and the implant. This, plus the aging of the implant material resulting from the inevitable breakdown of the molecular structure (another aspect of the Second Law), will, sooner or later, result in breaching of the implant surface. When this happens, the implant behaves as an open system and so will exchange mass with its environment, the body. Anyone contemplating using implants would benefit from a critical thinking assessment that would weigh the consequences of this inevitability against the perceived rewards.

Electric cars

Synopsis

One of the most widely publicized environmental issues in the Los Angeles basin in recent years is air pollution. Vehicles cause much, but hardly all, of the pollution. To cut back on this, California passed a "zero tail pipe emissions" law, which mandates a schedule for increasing sales of electrically powered cars. By 2003, California will require 10 percent of all cars sold in the state to have zero tail pipe emissions. In support of this, several of the major automakers have announced plans to mass-produce state-of-the-art electric cars.

As a result, political and media hype on electric cars has reached preposterous levels, particularly in California. I have collected newspaper and magazine clippings that quote supporters of this measure as claiming: electric cars are both environmentally responsible, since they do not produce emissions; and essentially free to operate, since all one has to do to "fill it up" is plug the batteries into any standard electrical outlet. Industrial sources claim that technology will soon produce super efficient batteries, which will double the range of existing lead acid batteries. In short, science and technology will produce battery-powered vehicles that will be nonpolluting and efficient while producing quiet and speedy transportation.

Critique

There are three issues here. First, zero emissions legislation directly contradicts the Second Law and ignores the First Law. Moving vehicles without generating pollution simply is not possible. In the case of electric cars, the obvious pollution products are removed from tailpipes and released at the power plants where the electricity is generated. Hence air pollution is concentrated at plants instead of being dispersed over the California freeway system. Furthermore, citizens will quickly find that charging their electric cars is not the same as turning on a lamp in their homes. It will take considerable energy, presumably in the form of hydrocarbons burned at power stations, to move these vehicles. As the First Law states, it takes energy to do useful work and moving cars on freeways requires substantial energy.

The second issue is that the media and political establishments have ignored the scientific protocol by relying on proprietary data and reports that have not undergone the scrutiny of peer review in the technical literature. The little data that has been published in the refereed literature raises serious questions about the true technical performance of electric vehicles and the economic and environmental consequences of their widespread use. Electric car proponents claim that these vehicles are substantially more efficient than conventionally powered vehicles. Thus, the total impact of widespread use of electric cars will be to drastically reduce societal energy demands. This is difficult to document, since a true measure of efficiency must consider the vehicle payload, driving conditions such as ambient temperature and acceleration cycles, and energy lost in transmission lines. Current lead acid batteries may weigh more than 1,100 pounds, and present car designs call for only two passengers. I have not located a refereed article examining all aspects of their efficiency. The ones I have seen have focused on performance as a function of ambient temperatures and acceleration cycles. Even though limited in scope, these studies raise serious questions as to whether this technology is ready for widespread use in public transportation.

Lastly, advocates of electric cars have ignored one consequence of their widespread use. Curiously, the same type of issue was ignored by the nuclear power industry in its growth years. The conventional lead acid battery has a lifetime of about 25,000 miles. What do we do with the used batteries? If these vehicles are widely used in the 21st century, then we need to develop long range plans and policies to dispose of the lead acid so as not to pass on another environmental crisis to subsequent generations. Some electric car advocates claim that many of the technical problems such as limited lifetimes, range, and limited acceleration capabilities will be solved by the next generation batteries. Well folks, they are talking about making these batteries out of stuff that will make lead acid look like tomato juice. From the thermodynamic perspective, energy used in transportation is merely a conversion of energy to work. Hence, both of MN2L apply. So before there is a widespread government mandated shift to battery operated vehicles, it would be wise to address ahead of time environmental problems arising from the disposal of batteries.

It seems, then, that the original rationale for electric cars was based on hype and not on the most elementary of physical concepts. Electric vehicles have obvious attributes that favor their use in niche situations, such as forklifts in warehouses. It is even possible these vehicles may have a niche in certain commuting environments, such as short distances over even terrain in mild temperatures. Moreover, even if their technical capabilities have been misrepresented to the public, it may still be that they are appropriate for wide use in public transportation. The issue here is not whether to be pro- or anti-electric cars, it is simply an unbiased appraisal of their capabilities and liabilities based on the scientific protocol and not on political and media hype.

The concerns raised above indicate that we have not yet thought through adequately many technical issues arising from the widespread use of electric cars. Upon careful reflection, reasonable solutions for these issues may be found. This is most likely to occur in an environment where quality control is established though the scientific protocol and not through legislation and advertisement.

Science and the Judiciary

The origins of our judiciary system predate by several centuries the Industrial Revolution and the ascendancy of science and technology. This has caused confusion as to how to accommodate science and technology in the judiciary process. The record during the 20th century is particularly pathetic. Moreover, as exemplified by the Scopes trial in the first part of the 20th century, as well as recent high profile criminal cases, and the breast implant suits mentioned earlier, there is heightened political, media, and public interest in the impact of science and technology on the courts. The common practice has been for both sides to call "expert" witnesses who then introduce highly technical data and analyses. Various strategies ranging from pure charismatic presentations to blatant attempts to confuse rather than enlighten the court seem to be involved. "Expert" witnesses are not brought in to assist the court in learning the truth but as advocates for one or the other side. In high profile cases these witnesses are handsomely compensated with no personal risk. These two factors have spawned a cottage industry in expert testimony. All of this provides great entertainment for us circus spectators; however, science and technology should play a more fundamental role in judicial proceedings than the size of the entertainment budgets of the court antagonists.

Thanks in large part to the breast implant cases, this situation is changing. Associate Justice of the Supreme Court Steven Breyer noted in a recent article in *Science* (see Suggested Reading) several promising developments. Judge Weinstein of New York appointed an independent panel of scientists to advise the court on cases involving breast implants. Judge Rosen of Michigan appointed a scientist to testify for the court in a case challenging a law prohibiting partial-birth abortions. Judge Stearns of Massachusetts appointed a scientist to advise his court on the scientific significance of evidence in a genetic engineering patent case.

Make no mistake about it, scientific issues now permeate judicial proceedings and no doubt will become even more important in the future. As noted by Justice Breyer, courts now must deal with highly technical issues regarding DNA fingerprinting, drug safety, leakage from toxic waste sites, and harmful effects of chemicals in industrial products.

Of course, there is strong vocal opposition to any change in the status quo from many lawyers and the cottage industry. The lawyers may be forced to use testimony that may go against their cases and may not be able to mount a rebuttal if their material does not meet appropriate standards. Well, so much for the myth that as officers of the court, lawyers are duty bound to pursue truth and justice. The cottage industry opposes these developments because many will not qualify as expert witnesses if reviewed publications are set as a standard, and fees likely will be reduced substantially.

Now let us consider the role of science and technology in the judiciary from a broader perspective. Except for mathematics, science is the most neutral and objective of all intellectual disciplines. Impartiality and objectivity are also judicial standards. Consider the judicial standard in the United States, "one is presumed innocent until proven guilty." This standard could be phrased in terms of a null hypothesis that the accused is innocent. The trial ideally is designed to minimize the chance of convicting an innocent person, that is, committing a type I error or rejecting a true hypothesis of innocence. The price of this safeguard is a greater chance of committing a type II error or letting the guilty walk.

The image of the blindfolded Lady of Justice, holding balance scales, symbolizes unbiased judgements and a critical weighing of the evidence. But these symbols apply equally well to the scientific protocol and critical thinking. It seems sensible that the judiciary should also formally adopt something akin to the scientific protocol when admitting scientific and technical evidence. Evidence relying on science and technology should be based on peer-reviewed methods and available to both sides. Moreover, scientific and technical expert testimony should be admitted in the context of a "friend of the court" and not as an advocate.

The argument that such a procedure destroys the natural adversarial nature of trials simply does not apply. The adversarial character of trials is always conducted within a fabric of rules and procedures dealing with the admissibility of evidence. A non-advocacy role for science and technology simply establishes fairer standards of admissibility.

To be sure, there are many obstacles to incorporating sound science in judiciary proceedings. One obvious hurdle is that judges have no procedure for recognizing who are qualified to serve as friends to the court. A thornier issue is teaching the judiciary how to deal with the scientific protocol. As indicated above, the more recent the result, the more the uncertainty among scientists. There are legitimate disputes over experimental and analysis procedures. Resolution of these issues could take more time than the trial. Appropriate data needs to be available in timely fashion to all parties. Finally, the judiciary must accept the fact that objective scientists looking at the same data may well draw different conclusions. Uncertainty and ambiguity in the courtroom are the bases of countless novels and films. By now readers should understand that science and technology cannot absolutely remove these factors. They can, however, provide more sophisticated levels of uncertainty and ambiguity than are now typically found in the courtroom.

As with science, there are no magic bullet solutions to these issues. It would certainly help if all involved in the judiciary process, that is, lawyers, judges, and jury, had a solid understanding of the scientific protocol and MN2L.

Envoi

The follies discussed earlier were picked more or less at random simply to illustrate a fundamental problem. Unfortunately, I have a lengthy list of other candidate examples. Thus it would be wrong to interpret the cases cited above as isolated and assume that simple fixes are all that is needed. Rather, they should be construed in the same way as spotting a cockroach in the kitchen: they are not one time isolated events. The four follies are symptoms of a much deeper issue in society that will not be resolved quickly. The message for readers is: do not rely on the media or political establishments for scientific and technical guidance. Instead, readers should take the initiative to make their own independent judgements using critical thinking, focusing only on results that have appeared in the peer reviewed scientific journals, and that are in conformance with MN2L.

Two themes emerge from these follies. The first is an arrogant disregard or appalling ignorance of the practice of science; the scientific protocol. It is ironic that this protocol, so important in establishing the United States as the world's dominant economic power; and in determining the successes of WWII, the Cold War, and the Gulf War, has fallen into such low repute. Media and political leaders in earlier eras understood enough about science and technology to allow the fruits of basic research to be applied to the pressing issues of their times. Such wisdom and courage apparently is lacking today.

The other theme is ignorance or disregard of Mother Nature's dictums that energy and mass are conserved and that entropy is always produced when energy is converted to work. When assessing economic and environmental policies, the tendency is to focus on energy requirements and the functional outputs or work performed. These are First Law issues. But as I have tried to show, Mother Nature guarantees there will always be waste products. This is a Second Law issue. The tendency is to address this only after the fact. Frosch's studies, noted in Chapter 2, show the economic and environmental benefits when MN2L are used together.

Two concerns discussed in this chapter were the dissemination of unreliable reports of scientific results and the need to formulate and implement public policies that properly include science and technology. The counsel to readers is to concentrate on critical thinking, the scientific protocol, and MN2L. Should we expect the media and political establishments to follow this advice? Hopefully yes, but there seems to be no motivation for them to do so. Hence the words of a noted critic seem particularly appropriate:

"No matter how cynical you become, it's never enough to keep up."

Lily Tomlin

Suggested Reading

Readers interested in an excellent summary of the Michelson-Morley experiment and its impact on science should consult *Physics Today*, 40, 5, 1987. This issue contains five very readable articles on the experiment and its impact on relativity.

A recent report on geologic conditions at Yucca Mountain is given by:

B. Wernicke, J. L. Davis, R. A. Bennet, P. Elosequi, M. J. Abolins, R. J. Brady, H. A. House, N. A. Niemi, and J. K. Snow, "Anomalous Strain Accumulation in the Yucca Mountain Area, Nevada," Science, 279, 2096-2100, 27 March 1998. The authors use Global Positioning System surveys from 1991 to 1997 to show that the buildup of strain in this region may be at least an order of magnitude greater than previous estimates.

An interesting assessment of these findings is given in the same issue:

R. A. Kerr, "A Hint of Unrest at Yucca Mountain," *Science*, 279, 2040-2041, 27 March 1998.

Legitimate scientific and technical concerns unrelated to *The New York Times* articles continue to surface regarding the suitability of Yucca Mountain as a nuclear waste repository. See for example *Science*, 283, 1235-1237, 1999.

Excellent non-technical summaries of the Yucca Mountain controversy are in *Science*, 268, 1836, 1995; *Science*, 271, 1664, 1996 and *EOS*, 76, 25,252, 1995.

Those interested in more material in the subsection "Science and the Law" should start with the following four references.

M. Angell, *Science on Trial*, W. W. Norton, 1997. Dr. Angell was the executive editor of the *New England Journal of Medicine* during the period when silicone breast implant cases were going to trial. This book provides the best synopsis to date of events during this period. The author points out the lack of refereed scientific literature in the evidence supporting the cause of the litigants. She also provides keen insight into the legal maneuvering, the dearth of scientific input into the litigation, and several of the facets of the anti-science movement in this country. Dr. Angell, perhaps more than anyone else, deserves credit for bringing to the public's attention the abuse of science in the courtroom.

S. Breyer, "The Interdependence of Science and the Law," *Science*, 280, 537-538, 24 April 1998. This is an important statement on the emerging and proper role of science in the courtroom. It is an encouraging sign that the author, an Associate Justice of the Supreme Court, published these views in *Science*.

E. Marshal, "New York Courts Seek 'Neutral' Experts," *Science*, 272, 12 April, 1996. This short article provides more detail on the appointment of a panel, which includes one scientist, by Judge Weinstein, the chief judge for the federal court of eastern New York, to advise the court on technical issues related to breast implant litigation.

A recent summary of the silicone breast implant litigation is given in an article by James T. Rosenbaum in *Science*, 276, 1524, 1997.

J. Harr, *A Civil Action*, Random House, 1996. This book describes, in detail, a famous case in Massachusetts in which several large companies were sued for polluting the groundwater in a community outside Boston. This pollution resulted in an outbreak of fatal cancers in the community. In contrast to silicone breast implants, a smoking gun agent was identified. The settlement reached in this litigation was the largest of its kind up to that time. The book gives an excellent account of what really happens with expert witness testimony in a highly charged and visible court case.

Two articles on the environmental impact of electrical cars are given in:

L. B. Lave, C. T. Hendrickson, and F. C. McMichael, "Environmental Implications of Electric Cars," *Science*, 268, 993-995, 19 May 1995.

"Downer for Electric Cars, Random Samples," *Science*, 274, 183-184, 11 October 1996.

"Responses to Iben Browning's Prediction of a 1990 New Madrid, Missouri, Earthquake." U. S. Geological Survey Circular 1083, 1993. This is a thorough and easy to read account of the public hysteria surrounding the false predictions of the earthquake in the New Madrid seismic zone. Appendix C of this volume provides a sample of the newspaper articles generated during this period.

Index